IET MATERIALS, CIRCUITS AND DEVICES SERIES 70

Self-Healing Materials

Other volumes in this series:

Self-Healing Materials

From fundamental concepts to advanced space and electronics applications
2nd Edition

Brahim Aïssa, Emile Haddad
and Wes R. Jamroz

The Institution of Engineering and Technology

Published by The Institution of Engineering and Technology, London, United Kingdom

The Institution of Engineering and Technology is registered as a Charity in England & Wales (no. 211014) and Scotland (no. SC038698).

First published 2014

Second edition 2019

The Institution of Engineering and Technology
Michael Faraday House
Six Hills Way, Stevenage
Herts, SG1 2AY, United Kingdom

www.theiet.org

British Library Cataloguing in Publication Data
A catalogue record for this product is available from the British Library

ISBN 978-1-78561-992-2 (hardback)
ISBN 978-1-78561-993-9 (PDF)

Typeset in India by MPS Limited
Printed in the UK by CPI Group (UK) Ltd, Croydon

Contents

List of figures

List of tables

Preface

The development of self-healing materials has become a fast growing area of research for applications in future composite structures. Over the past few decades, there has been a huge interest in materials that can self-heal, as this property can increase materials' lifetime, reduce replacement costs and improve product safety. Self-healing systems can be made from a large variety of polymers and metallic materials.

This second edition of the book is dedicated to self-healing materials and their applications. Special emphasis has been put on the use of self-healing systems in space.

In addition to giving an overview of the major works in the field, the book also provides the introductory material. In this way, the book targets the readership among those unfamiliar with the subject as well as those among academia and industrial research community who are experts in the field. The book is based on expertise of the authors and their work carried out over the last 10 years.

Summary of the general concepts of various self-healing processes is detailed in Chapter 1. Chapters 2–5 are a review of the natural self-healed systems and those based on polymers and composites. Chapter 6 reviews the experimental results obtained on the advanced fabrication processes, while Chapter 7 details the self-healing systems for space applications. Chapter 8 revolves around the self-healing of composites materials subjected to hyper-velocity impacts simulating orbital debris while Chapter 9 deals with the mitigating the effect of space small debris on COPV in space with fiber sensors monitoring and self-repairing materials and Chapter 10 concludes with an outlook into the future developments and applications.

The book is amended with an extensive and up-dated survey of the published articles and conference reports. It is hoped that the book will be of interest to those involved in the investigation of self-healing materials at large, ranging from novice students to the most experienced end-users. The intention is also to stimulate debate and reinforce the importance of a multidisciplinary approach in this exciting field.

Acknowledgements

First of all, we would like to offer our special thanks to Ms. Jane Bachynski, President of MPB Communications Inc., for her continuous and valuable support that has created a magic environment where the business objectives are realised through the cultivation of innovation and by harnessing scientific curiosity.

This book would not have been possible without the tremendous support of many people. In particular, we sincerely express our gratitude to our colleagues from various institutions and organisations who provided us with assistance to our work. Especially we thank Darius Nikanpour, Stéphane Gendron and Philip Melanson from the Canadian Space Agency; Mohamad Asgar and Suong V. Hoa from Concordia University; Hasna Hena Zamal from INRS; Daniel Therriault from Ecole Polytechnique de Montreal; Mourad Nedil from UQAT; Philippe G. Merle from the Department of National Defense; Jason Loiseau, Jimmy Verreault and Andrew Higgins from McGill university; Christopher Semprimoschnig, Iain McKenzie, Nikos Karafolas and Philippe Poinas from the European Space Agency (ESA).

We wish to acknowledge the help provided by Dr. Roman Kruzelecky, Kamel Tagziria and Roy Josephs from MPB Communications Inc. for their assistance with experimental data.

We wish to underline the valuable help provided by the staff of The Institution of Engineering and Technology (IET) publishing group, with a special mention to Sarah Lynch, Olivia Wilkins and Paul Deards for their availability and willingness to provide guidance and advice during the preparation of the manuscript.

We would like to gratefully acknowledge the financial and technical contribution of the Canadian Space Agency, the Natural Science and Engineering Research Council (NSERC) of Canada and the European Space Agency. A special thank goes as well to Qatar Environment and Energy Research Institute (Hamad Bin Khalifa University, Qatar Foundation).

List of abbreviations

ASTM	American Society for Testing and Materials
AH	aluminium honycomb
AHPCS	allylhydrid-opoly-carbosilane
AO	atomic oxygen
AFM	atomic force microscopy
AHPCS	allylhydrid- opoly-carbosilane
AG	as grown
BGE	1-butyl glycidyl ether
CVD	chemical vapour deposition
CLEO	circular low earth orbit
CMC	ceramic matrix composite
COPV	composite overwrapped pressure vessels
CFRP	carbon fibre reinforced polymers
CTE	coefficient of temperature expansion
CAI	compression after impact
CWL	centre wavelength
CL	cross linking agent
CNT	carbon nanotube
DCPD	dicyclopentadiene
DETA	diethylenetriamine
DGEBPA	diglycidyl ether of bisphenol A
DNA	deoxyribonucleic acid
DCM	dichloromethane
EMA	ethylenemaleic anhydride
EMMA	poly-(ethylene-co-methacrylic acid)
EDS	energy dispersive
ECR	electron cyclotron resonance

FBG	fibre Bragg grating
FRP	fibre-reinforced polymer
FEA	finite element analysis
GFRP	glass fibre reinforced polymers
GEO	geostationary orbit
HEO	highly elliptical orbit
IFEN	in-flight entertainment network
IR	infrared
ISS	International Space Station
LN2	liquid nitrogen
LEO	low earth orbital
MMOD	micrometeorites and orbital debris
MISSE	Materials International Space Station Experiment
MLIB	multi-layer insulation blankets
MWCNT	multi-walled carbon nanotube
MD	molecular dynamics
MUF	melamine-urea-formaldehyde
MTS	material testing system
NOAX	non-oxide adhesive experiment
NCA	5-norbornene-2-carboxylic acid
NPs	nanoparticles
NDM	non-destructive methods
PLD	pulsed laser deposition
PCy3	tricyclohexylphosphine
PMU	polyoxymethylene urea
PMUF	poly melamine urea-formaldehyde
PUF	poly urea-formaldehyde
PDMS	polydimethylsiloxane
PMMA	poly(methyl methacrylate)
PMUF	poly melamine-urea-formaldehyde
PDM	special purpose dexterous manipulator
PC	polycarbonate
PF	purified

PVD	pulsed vapour deposition
POSS	polyhedral oligomeric silsesquioxane
RGC	ruthenium Grubbs' catalyst
ROMP	ring opening metathesis polymerisation
RPM	revolutions per minute
RBM	radial breathing mode
SWCNT	single-walled carbon nanotube
SOFC	solid oxide fuel cells
SEM	scanning electron microscopy
SMA	shape-memory alloy
TransHab	transit habitat
TEM	tunnelling electron microscopy
TGA	thermo-gravimetry analysis
TOF	turn over frequency (to measure a ROMP efficiency)
TON	turn over number (to measure a ROMP efficiency)
TDCB	tapered double cantilever beam
Tg	glass transition temperature
UV	ultraviolet
PDMS	polydimethylsiloxane(s)
UF	urea-formaldehyde
UHV	ultrahigh vacuum
VUV	vacuum ultraviolet radiation
WTDCB	wide tapered double cantilever beam
XPS	X-ray photoelectron spectroscopy
XRD	X-ray diffractometer
YSZ	yttria-stabilised-zirconia
ENB	5-ethylidene-2-norbornene
2D	two dimensional

Chapter 1

Introduction

A major challenge for space missions is that all materials degrade over time and are subject to wear, especially under extreme environments and external impacts. Micrometeoroids and orbital debris, particularly in the lower earth orbit, present a continuous hazard to orbiting satellites, spacecrafts and the international space station. Space debris includes all nonfunctional, man-made objects and fragments. As the volume of debris continues to grow, the probability of collisions that could lead to potential damage will consequently increase. Thus, impact events are inevitable during the lifetime of a space structure, and once they are damaged they are hardly repairable.

A unique challenge of space missions is the intrinsic requirement for a robust design that does not rely on in-orbit repair. Such repairs or service activities such as the ones carried out for the Hubble Space Telescope or during the return to flight campaign of NASA in 2005 (Mission STS-114) are exceptions. They require astronauts, extensive preparation and are very costly even if performed by space robots. The long-duration missions together with the expansion of both spacecraft and human-explorers sizes are the main reasons behind the escalation of the recorded failure-tolerant space-materials. In many of these missions, repair of materials will be virtually impossible.

In the last two decades, advances in science and technology have led to the creation of materials with properties so unique that they offer applications previously unheard of. These novel materials and their implementation have formed a new branch of material science: self-repairing.

In addition, the self-repair is functional during the qualification tests on the ground which increases the reliability and the ratio of components that successfully pass these tests. Even small cracks that develop during the harsh conditions of the launch could be repaired later during the flight.

Polymeric composites that are used in a variety of space applications are susceptible to damage, which is induced by mechanical, chemical, thermal, ultraviolet radiation or a combination of these factors [1]. When the polymer composites used as structural materials become damaged, there are only a few methods available to attempt to extend their functional lifetime. Materials' failure normally starts at the nanoscale level and is then amplified to the micro- up to the macroscale until a catastrophic failure occurs. The ideal solution would be to block and eliminate damage as it occurs at the nano/microscale and restore the original material properties.

Ideal repair methods are ones that can be executed quickly and effectively directly on the damaged site, thereby eliminating the need to remove a component for repair. However, the mode of damage must also be taken into consideration as repair strategies that work well for one mode might be completely useless for another. For example, matrix cracking can be repaired by sealing the crack with resin, whereas fibre breakage would require replacement with new fibres or a fabric patch to achieve recovery of strength [2,3].

Since the damage deep inside materials are difficult to perceive and repair, it would be better to have materials with intrinsic self-repair capabilities – a sort of biomimetic healing functionality. Indeed, self-healing materials could be the appropriate solution. However, this approach is not easy to make on place because of the harsh conditions characterizing space environment.

Many naturally occurring parts in animals and plants are provided with such a self-healing function [4–8]. In the case of the healing of a skin wound, for example, the defect is temporarily plugged with a fibrin clot, which is infiltrated by inflammatory cells, fibroblasts and a dense capillary plexus of new granulation tissue. Subsequently, proliferation of fibroblasts with new collagen synthesis and tissue remodelling of the scar become the key steps.

Similar processes take place in the healing of a broken bone. The healing process includes internal bleeding forming a fibrin clot, development of an unorganised fibre mesh, calcification of fibrous cartilage and conversion of calcification into fibrous bone and lamellar bone. Clearly, the natural healing in living bodies depends on rapid transportation of repair substance to the injured part and reconstruction of the tissues. Having been inspired by these findings, continuous efforts are now being made to mimic natural materials and to integrate self-healing capability into polymers and polymer composites [5–7]. Thus, self-healing materials exhibit the ability to repair themselves and to recover functionality using the resources inherently available to them. Whether the repair process is autonomic or externally assisted (e.g., by heating), the recovery process is triggered by damage to the material. Self-healing materials offer a new route towards safer, longer-lasting products and components. The progress has opened the new field of intelligent materials [9].

One of the earlier healing methods for fractured surfaces was 'hot plate' welding, where polymer pieces were brought into contact above the glass transition temperature of the material, and this contact was maintained long enough for interdiffusion across the crack face to occur and restore strength to the material. It has been shown, however, that the location of the weld remains the weakest point in the material and thus, the favourable site for future damage to occur [10,11].

For laminate composites, resin injection is often used to repair damage which occurs in the form of delamination. This can be problematic, however, if the crack is not easily accessible for such an injection. For fibre breakage in a laminate composite, a reinforcing patch is often used to restore some of the strength to the material. Often, a reinforcing patch is used in conjunction with resin injection to restore the greatest amount of strength possible [12]. None of these methods of repair is an ideal solution to damage in a structural composite material. These methods are temporary solutions to prolong the lifetime of the material and each of these repair strategies requires monitoring of the damage and manual intervention. This greatly increases the cost of the material by requiring regular maintenance and service. Alternative healing strategies are therefore of great interest. Moreover, with polymers and composites being increasingly used in structural applications in space, automobile, defence and construction industries, several techniques have been developed and adopted for repairing visible or detectable damages. However, these conventional repair methods are not effective, for example, for healing invisible microcracks within the structure during its service life. In response, the concept of 'self-healing' polymeric materials was proposed in the 1980s [13] as a means of healing invisible microcracks. Early researches by Dry and Sottos [14] published in 1993 and then White *et al.* [9] in 2001, further inspired interest in these materials and their applications. Examples of such interest were demonstrated by the United States Air Force [15] and the European Space Agency [16] investments in self-healing polymers.

Conceptually, self-healing materials have the built-in capability to substantially recover their mechanical properties after damage. Such recovery can occur autonomously and/or be activated after an application of a specific stimulus (e.g., heat, radiation, pressure and so on). As such, these materials are expected to contribute greatly to the safety and durability of polymeric components without the high costs of active monitoring or external repair. Throughout the development of this new range of smart materials, the mimicking of biological systems has been used as a source of inspiration [17].

Typical examples of self-healing materials can be found in polymers, metals, ceramics and their composites, which are subjected to a wide variety of healing principles. Healing can be initiated by means of an external source of energy as was shown in the case of a bullet penetration [18] where the ballistic impact caused local heating of the material by allowing self-healing of ionomers. In the case of self-healing paints used in the automotive industry, small scratches can be restored by solar heating [19]. Single cracks formed in polymethyl methacrylate specimens at room temperature could also be completely restored if the temperature was raised above the glass transition temperature [10,13,20]. The presence of noncovalent hydrogen bonds [21] in mechano-sensitive polymers can allow a rearrangement of principal chemical bonds so that they can be used for self-healing. However, noncovalent processes may limit the long-term stability in structural materials. Force-induced covalent bonds can be activated by incorporating mechanophores (mechanically sensitive chemical groups) in polymer strands [22]. Numerical studies have also shown that nanoscopic gel particles, which are interconnected in a macroscopic network by means of stable and labile bonds, have the potential to be used in self-healing applications. Upon mechanical loading, the labile bonds break and bond again with other active groups [23]. Contact methods have also been investigated where self-healing of the damaged samples is activated with a sintering process, which increases the contact adhesion between particles [24]. Although these approaches are quite interesting, it is believed that the most promising methods for self-healing applications involve the use of nano/microparticles [25], hollow tubes and fibres [26], microcapsules [9], nanocontainers [27] or microfluidic vascular systems [28], filled with a fluid healing agent (e.g., epoxy for composite materials [29], corrosion inhibitor for coatings [30] and so on) and dispersed in the host material. When the surrounding environment undergoes changes such as temperature, pH, cracks or impacts, then the healing agent is released.

The efficacy of healing is therefore regulated by the balance of the rate of damage versus the rate of healing. The rate of damage for a material is dictated by external factors such as the frequency of loading, strain rate and stress amplitude. However, the rate of healing can be tailored or tuned to specific damage modes by, for example, varying the reaction kinetics through species concentration or temperature. Thus, the goal of self-healing is to achieve material stasis by balancing the rate of healing and the rate of damage.

The number of publications dealing with various aspects of self-healing materials has increased markedly in recent years. Figure 1.1 shows how the number of refereed various articles in self-healing field has steadily increased since 2001, based on data collected from the Engineering Village [31]. Along with the increase in the number of publications in this area comes a need for a comprehensive review, and the objective of this book is to address this need.

In addition, the vast majority of the surveyed articles deal with polymer composites. Due to the large number of articles involved and the lack of electronic access to many conference proceedings, the emphasis of this review is on the more accessible refereed journals. It was not practical to cover all of the published articles, therefore, an attempt was made to select representative articles in each of the relevant categories.

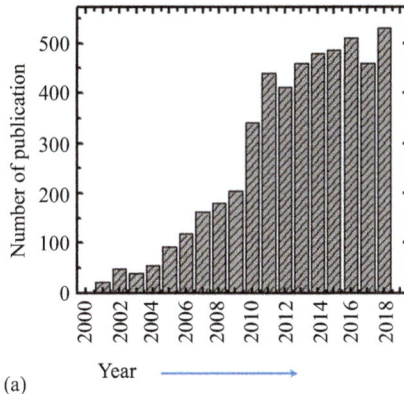

(a)

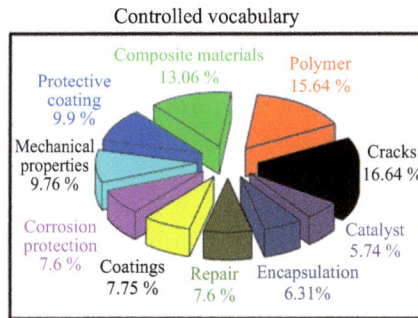

(b)

Figure 1.1   *(a) Recent refereed publications related to the field of self-healing materials and (b) the distribution of the employed keywords vocabulary. All published languages were included, all documents types, including journal and conferences articles, report review, conference proceeding and monograph published chapters were recorded. Statistics are available from 2000 to 2013 inclusively.*
[Adapted from Engineering Village web-based information service © Elsevier BV [31]]

On the whole, research in this field is still in its infancy. More and more scientists and companies are interested in different aspects of the topic. Innovative measures and new knowledge of the related mechanisms are constantly emerging. Depending on the method of healing, self-healing polymers and polymer composites can be classified into two categories: (i) intrinsic ones that are able to heal cracks by the polymers themselves, and (ii) extrinsic in which healing agent has to be pre-embedded.

This book begins with a description of the methods for evaluating different self-healing approaches efficiencies. Next some examples of different approaches proposed to heal the thermoplastic systems are discussed and this is followed by covering the preparation and characterisation of self-healing of thermosetting systems. A particular emphasis is given to space applications. The book concludes by considering future research.

The book will be of interest to those unfamiliar with the subject as well as to those among the academic and industrial research community who are currently working in the field.

The book is based on the unique expertise of the authors and the pioneering work carried out over the last 10 years.

References

[1]   C.B. Bucknall, I.C. Drinkwater and G.R. Smith, *Polymer Engineering and Science*, 1980, **20**, 6, 432.

[2]   B. Aïssa, D. Therriault, E. Haddad and W. Jamroz, "Self-Healing Materials Systems: Overview of Major Approaches and Recent Developed Technologies", *Advances in Materials Science and Engineering*, 2012, **2012**, Article ID 854203.

[3]   E. Haddad, R.V. Kruzelecky, W.P. Liu and S.V. Hoa, *Innovative Self-repairing of Space CFRP Structures and Kapton Membranes: A Step Towards Completely Autonomous Health Monitoring & Self-healing*, Final Report, Contract No: CSA 28-7005715, Canadian Space Agency, Saint-Hubert, Quebec, Canada, 2009.

[4]   R.S. Trask, H.R. Williams and I.P. Bond, *Bioinspiration and Biomimetics*, 2007, **2**, 1, P1.

[5]   G.W. Hastings and F.A. Mahmud, *Journal of Intelligent Material Systems and Structures*, 1993, **4**, 4, 452.

[6]   P. Martin, *Science*, 1997, **276**, 5309, 75.

[7]   A.I. Caplan in *Ciba Foundation Symposium 136 – Cell and Molecular Biology of Vertebrate Hard Tissues*, Eds., D. Evered and S. Harnett, Wiley, New York, NY, 1988, p. 3.

[8]   S.F. Albert and E. Wong, *Clinics in Podiatric Medicine and Surgery*, 1991, **8**, 4, 923.

[9]   S.R. White, N.R. Sottos, P.H. Geubelle, *et al.*, *Nature*, 2001, **409**, 6822, 794.

[10] H.H. Kausch, *Pure and Applied Chemistry*, 1983, **55**, 5, 833.

[11] D. Liu, C.Y. Lee and X. Lu, *Journal of Composite Materials*, 1993, **27**, 13, 1257.

[12] T.A. Osswald and G. Menges, *Materials Science of Polymers for Engineers*, Hanser Publishers, Munich, Germany, 2003.

[13]   K. Jud, H.H. Kausch and J.G. Williams, *Journal of Materials Science*, 1981, **16**, 1, 204.

[14]   C.M. Dry and N.R. Sottos, 'Passive smart self-repair in polymer matrix composites', *SPIE Proceedings Volume 1916, Smart Structures and Materials: Smart Materials*, Bellingham, Bellingham, WA, 1993, p. 438.

[15]   H.C. Carlson and K.C. Goretta, *Materials Science and Engineering B: Solid-state Materials for Advanced Technology*, 2006, **132**, 1–2, 2.

[16]   C. Semprimoschnig, Enabling Self-healing Capabilities – A Small Step to Bio-mimetic Materials, European Space Agency Materials, Report Number 4476, European Space Agency, Noordwijk, The Netherlands, 2006.

[17]   S. Varghese, A. Lele and R. Mashelkar, *Journal of Polymer Science, Part A: Polymer Chemistry Edition*, 2006, **44**, 1, 666.

[18]   R.J. Varley and S. van der Zwaag, *Acta Materialia*, 2008, **56**, 19, 5737.

[19]   S. Van der Zwaag, Ed., *An Introduction to Material Design Principles: Damage Prevention versus Damage Management Self-Healing Materials*, Springer, Dordrecht and Noordwijk, The Netherlands, 2007.

[20]   H.H. Kausch and K. Jud, *Plastics and Rubber Processing and Applications*, 1982, **2**, 3, 265.

[21]   R.P. Sijbesma, F.H. Beijer, L. Brunsveld, *et al.*, *Science*, 1997, **278**, 5343, 1601.

[22]   D.A. Davis, A. Hamilton, J. Yang, *et al.*, *Nature*, 2009, **459**, 7243, 68.

[23]   G.V. Kolmakov, K. Matyjaszewski and A.C. Balazs, *ACS Nano*, 2009, **3**, 4, 885.

[24]   S. Luding and A.S.J. Suiker, *Philosophical Magazine*, 2008, **88**, 28–29, 3445.

[25]   M. Zako and N. Takano, *Journal of Intelligent Material Systems and Structures*, 1999, **10**, 10, 836.

[26]   S.M. Bleay, C.B. Loader, V.J. Hawyes, L. Humberstone and P.T. Curtis, *Composites Part A: Applied Science and Manufacturing*, 2001, **32**, 12, 1767.

[27]   D.G. Shchukin and H. Möhwald, *Small*, 2007, **3**, 6, 926.

[28]   K.S. Toohey, N.R. Sottos, J.A. Lewis, J.S. Moore and S.R. White, *Nature Materials*, 2007, **6**, 8, 581.

[29]   T. Yin, L. Zhou, M.Z. Rong and M.Q. Zhang, *Smart Materials and Structures*, 2008, **17**, 1, 015019.

[30]   S.H. Cho, S.R. White and P.V. Braun, *Advanced Materials*, 2009, **21**, 6, 645.

[31]   Engineering Village, Elsevier BV, The Netherlands, www.Elsevier.com/online-tools/engineering-village.

Chapter 2
Natural systems and processes

Natural systems, such as biological materials, have the ability to sense, react, regulate, grow, regenerate and heal themselves. Biological materials constitute most of the bodies of plants and animals around us. They allow cells to function, eyes to capture and interpret light, plants to respond to the light and animals to move or fly. This multitude of operations has always inspired mankind to make materials and devices, which simplify many of our day-to-day functions. One remarkable property of natural materials and structures is their ability of self-sealing and self-healing. Many animals and plants regenerate tissues or even whole organs after injury. However, biological repair processes are generally very complex and an adaption into a technical system is not easy. Recent advances in chemistry and micro- and nanoscale fabrication techniques have enabled biologically inspired technical systems that mimic many of these remarkable functions. For example, self-cleaning surfaces are based on the super-hydrophobic effect, which causes water droplets to roll off with ease, carrying away dirt and debris. Design of these surfaces is based on the hydrophobic micro- and nanostructures of a lotus leaf.

In this chapter, the most successful strategies are examined, and future research directions, opportunities and outlooks are discussed. As example, the self-cleaning phenomenon and the self-healing of human wound are described in detail.

2.1 Introduction

The study of biological systems as structures dates back to the early twentieth century. The classic work by D'Arcy W. Thompson [1], first published in 1917, can be considered the major pioneer work in this field. He looked at biological systems as engineering structures and formulated the relationships that described their form. In the 1970s, Currey investigated a broad variety of mineralised biological materials and wrote the well-known book *Bone* [2]. Another important work is Vincent's *Structural Biomaterials* [3].

Indeed, biological structures are a constant source of inspiration for solving a variety of technical challenges in architecture [4], aerodynamics and mechanical engineering [5,6], as well as in materials science [7]. Natural materials consist of relatively few constituent elements, which are used to synthesise a variety of polymers and minerals. On the contrary, the history of human inventions is characterised by the use of many elements. This led to the invention of materials with special properties, which are not found in nature. The ages of copper, bronze and iron were later followed by the industrial revolution based on steel and the information age based on silicon semiconductors. All these materials require high temperatures for fabrication and biological organisms have no access to them. Nevertheless, nature has developed, with comparatively few base substances, a range of materials with remarkable functional properties. The key is a complex, often hierarchical, structuring of the natural materials [7–9], which results from the fact that they grow according to a recipe stored in the genes, rather than being fabricated according to an exact design.

This is why the design strategies of biological structures are not immediately applicable to the design of new engineering materials (Table 2.1) [10]. The first major difference is in the range of choice of elements, which is far greater for the engineer. Elements such as iron, chromium and nickel are very rare in biological tissues and certainly not used in a metallic form, as would be the case of steel. Iron is found in red blood cells, for instance, as an ion bound to the protein haemoglobin and its function is certainly not mechanical but chemical. Most of the structural materials used by nature are polymers or composites of polymers and ceramic particles. Such materials would generally not be the first choice of an engineer to build strong and long-lasting mechanical structures. Nevertheless, nature uses them to build trees and skeletons. The second major difference is the way in which materials are made. While the engineer selects a material to fabricate a part according to an exact design, nature goes the opposite way and grows both the material and the whole organism (a plant or an animal) using the principles of biologically controlled self-assembly. This provides control over the structure of the material at all levels of hierarchy and is certainly a key to the successful use of polymers and composites as structural materials.

Bio-inspiration is not just a consequence of an observation of naturally occurring structures. The reason is that nature has a multitude of boundary conditions which we do not know *a priori* and which might all be of great importance for the development of a specific structure. In summary, biological and engineering materials are driven by a very different choice of base components and a different mode of fabrication. As a result, different

Table 2.1 Examples of biological versus engineering materials

Biological material	Engineering material	References
Light and soft elements dominate: H, C, N, O, Ca, P, S, Si, ...	Large variety of elements: Co, Ni, Fe, Cr, Ag, Al, Si, C, N, O, ...	[7–10]
The growth process is controlled by biological self-assembly (approximate design)	Fabrication from melts, powders, solutions and so on (exact design)	[8,10]
Hierarchical structuring is respected at all size levels	Forming of the part and micro- and nanostructuring of the material	[8,10]
Adaptation of form and structure to the function	Selection of material according to function	[7–10]
Modelling and remodelling: capability of adaptation to changing environmental conditions	Secure design (considering possible maximum loads as well as fatigue)	[10]
Healing: capability of self-repair		[10]
Molecular self-assembly	Particles self-assembly (large group of particles that self-assemble into thermodynamically stable system)	
Cleaning: capability of self-clean	Use the photo-catalytic effect of large band gap materials that decomposes pollutant using UV of the sunlight	[44,45]
Photosynthesis: making glucose from CO_2 and H_2O molecules through sunlight	Photovoltaic system collects energy from the sun	
Sticking property (plants sticking to clothes)	Velcro materials	[11]

strategies have to be implemented to achieve the desired performance (see Table 2.1).

Building a complete engineering system similar to nature requires a careful study of the biological system and an understanding of the structure–function relationship of the biological material. Such an approach must be considered in the context of the existing physical and biological constraints. When designing biomimetic materials, we often inspired by copying the nature, with different degrees of complexity. The simplest level is to focus on one *natural* function and develop the engineered material to perform exclusively this specified function. Numbers of materials are now being developed to be implemented to this view; however, many technical challenges need to be addressed before thinking about the commercial level. A first successful example is the development of the Velcro materials. In 1948, the Swiss engineer George de Mestral started to observe the sticking effect of the burrs of the burdock plant to his clothes. It took a few years (1955) to patent a product and up to 1959 to have a commercial product [11].

Another successful example is to use microcapsules with a healing agent and a catalyst to repair cracks in composite materials. These processes, based on the metathesis reaction, were first demonstrated by Chauvin *et al.* in 1971 [12]; however, the proper catalyst was missing. It took until 1990 before Schrock could develop the first catalyst working in a vacuum, and a few years more for Grubbs to develop a commercial metal compound catalyst that is stable in air and allowed the production of the final product. Metathesis is now largely used in the development of pharmaceutical and advanced plastic materials [12].

2.2   Growth and functional adaptation

Growth is a sensitive process that can be influenced by many external conditions including temperature, mechanical loading and by supply of light, water or nutrition [1]. A living organism must necessarily possess the ability of adaptation to external needs, while possible external influences on a technical system must be typically anticipated in its design. This often leads to considerable 'over-design' (Figure 2.1). This aspect of functional adaptation is particularly fascinating for the materials scientist since solutions already discovered by nature can serve as sources of inspiration. The subject was pioneered by Thompson whose classical book in 1919 *On Growth and Form* was republished several times [1]. This early text mostly relates the 'form' (or shape) of biological objects to their function. Even earlier, the relationships between anatomy (i.e., structure) and function of living systems had been explored by Leonardo da Vinci and Galileo Galilee [10]. The latter is often considered as the father of biomechanics. Among his many discoveries, he recognised that the shape of an animal's bones is to some extent adapted to its weight. Long bones of larger animals typically have a smaller aspect ratio.

Galileo's explanation is a simple scaling argument, based on the fact that the weight of an animal scales with the third power of its linear dimension, while the structural strength of its bones scales with its cross-section, that is, the square of the linear dimension. Thus, the aspect ratio of long bones has to decrease with the body weight of the animal. This is also a good example of functional adaptation.

Different strategies in designing a material result from the two paradigms of 'growth' and 'fabrication' (Table 2.1). For engineering materials, a machine part is designed and then the material is selected according to the knowledge and experience regarding the functional requirements, taking

*Figure 2.1   (a) Schematic of a drop of water on a lotus leaf, (b) the
microstructure of a leaf, (c) microscopic and (d) macroscopic
schematic views of a butterfly wing*

into account possible changes in those requirements during service (e.g.,
typical or maximum loads) and fatigue (and other lifetime issues) of the
material. The strategy is static, as the design is made at the beginning and
must satisfy all needs during the lifetime. The fact that natural materials are
growing rather than being fabricated leads to the possibility of a dynamic
strategy: it is not the exact design of the organ that is stored in the genes, but
rather a recipe to build it. This means that the final result is obtained by an
algorithm rather than by the replication of a design. The advantage of this
approach is that it allows flexibility at all levels. First, it permits adaptation
to the function, while the body is growing. For example, a branch growing
in the direction of the wind may grow differently than a branch that faces in
the opposite direction, without any change in the genetic code. Second, it
allows the growth of hierarchical materials, where the microstructure at
each position of the part is adapted to the local needs [13]. This is linked to
the idea of robustness: nature has evolved structures that are capable of
surviving/withstanding/adapting to a range of different environments, while
man-made materials are generally less flexible in their use.

Adapting the form (of a whole part or organ, such as a branch or a vertebra) is the first aspect of functional adaptation. A second possibility, which relates more directly to materials science, is the functional adaptation of the microstructure of the material itself (such as the wood in the branch or the bone in the vertebra).

This dual need for optimisation of the part's form and the material's microstructure is well known for any engineering problem. However, in natural materials, shape and microstructure become intimately related due to their common origin, which is the growth of the organ. This aspect has been discussed in detail by Jeronimidis in his introductory chapters to a book on *Structural Biological Materials* [13]. Growth implies that 'form' and 'microstructure' are created in the same process but in a stepwise manner. The shape of a branch is created by the assembly of molecules into cells, and cells into wood with a specific shape. Thus, at every size, the branch is both form and material: the structure becomes hierarchical.

A small step in the growth in biology that probably is easier to define is the organ regeneration plants and animals. In 2012, a conference has been held to find possible signatures of the plants and animals organ regeneration [14] and to look at the reasons behind the fact that many human tissues have such a limited capacity to regenerate. This was an exercise in trying to find out whether the ability has been lost in our evolution, or if it rather never existed. Table 2.2 provides a summary of regeneration in nature based on the work of [14,15].

Table 2.2 Summary of organ regeneration in nature

Biological entity	Regeneration
Arabidopsis (small flowering plant related to cabbage)	The *Arabidopsis thaliana* species is one of the model organisms largely used for studying plant biology and the first plant to have its entire genome sequenced. New cells emerge from a stem cell niche, in the root. After removal of the stem cell compartment, the root retains the ability to rapidly regenerate, presumably by regenerating the stem cell niche from differentiating cells. The National Aeronautics and Space Administration plans to grow Arabidopsis on the moon in 2015 in the 'LPX First flight of Lunar plant growth experiment' [16]
Annelid worms	They have the capacity to regenerate after amputation – all these animals reproduce by fission
Hydra (1 cm long, radial, small freshwater animal)	Mid-gastric amputation stimulates apoptosis in hydra polyps. These dying cells release signals, such as Wnt3, that stimulate the regenerative response
Zebrafish (4 cm)	In amputated larval zebrafish fins, the regeneration of somatosensory neurons in the skin is guided by hydrogen peroxide. Heart regeneration in zebrafish is accompanied with an increase of localised production of hydrogen peroxide by epicardial cells in stimulating cardiomyocyte proliferation during heart regeneration

(Continues)

Table 2.2   (Continued)

Biological entity	Regeneration
Axolotl (Mexican walking fish or Mexican salamander, about 25 cm)	The spinal cord regeneration was tested using neural stem cells that can be isolated, cultured into neurospheres and transplanted to contribute multiple cell types to a regenerating spinal cord
Killifish (small fish, 5–15 cm)	High regenerative capacity – regenerates amputated fins like other teleost fish
Xenopus (aquatic frog)	Regains function after a severe spinal cord transection injury. The regeneration of amputated adult *Xenopus* limbs is enhanced with stem cells. A cut tail of *Xenopus laevis* regenerates
Newt (aquatic amphibian of the family *Salamandridae*)	The newts are capable of regenerating eye lenses and different types of neurons after an injury to the adult newt brain. The regeneration involves mainly the regeneration of lost dopaminergic neurons once the appropriate number of dopaminergic neurons is restored, the regeneration ceases. The regenerative capacity in newts is not altered by repeated regeneration and ageing
Salamander	Dopamine controls neurogenesis in the adult salamander midbrain in homeostasis and during regeneration of dopamine neurons
Acomys (mouse, African spiny mouse)	Partial regeneration – able to transplant hepatocytes, pancreatic beta cell islets and thymocytes into mouse lymph nodes, where these tissues grow, become vascularised and may have their own independent functionality. In particular in the African spiny mouse, the dermis and skin are extremely weak and tear easily away from their bodies, but the resultant large wounds are repaired remarkably well and regenerate hair follicles throughout the wound bed. Moreover, they can regenerate large hole punches in their ears much better than laboratory strains of mice. Thus, it is likely that unsuspected examples of regenerative capacity remain to be discovered, representing new model systems that will inform us of why regeneration does (or does not) occur
Human	Many human tissues have a limited capacity to regenerate

	The young human heart might have some capacity to regenerate after injury, a property thought to be absent in adult humans:
	
	• self-healing of the skin
	• self-healing of broken bones
	• partial regeneration of the liver, the concepts of regeneration using as a template the liver:
	  – local cells (either differentiated hepatocytes or tissue specific-stem cells) proliferate and repopulate the injured area, providing as well important factors for growth
	  – an artificial scaffold is seeded with cells to repopulate and reconstruct the lost part of the organ
	  – cells from bone marrow can be brought *via* blood vessels to the injured area and contribute to the repair or regeneration

Adapted and reproduced, with permission, from [14] and [15].

2.3   Hierarchical structuring

Hierarchical structuring is one of the consequences of the growth of organs. Examples for hierarchical biological materials are bones [17–20], trees [21–24], seashells [25], spider silk [26], the attachment systems of geckos [27], super-hydrophobic surfaces (lotus effect) [28], optical microstructures [28,29], the exoskeleton of arthropods [30,31] or the skeleton of glass sponges [32]. Hierarchical structuring allows for the construction of large and complex organs based on much smaller, often very similar, building blocks. Examples of such building blocks are collagen fibrils in bone, which have units with a few hundred-nanometre thickness and can be assembled into a variety of bones with very different functions [2,18,19,33]. Moreover, hierarchical structuring allows for the adaptation and optimisation of the material at each level of hierarchy to yield outstanding performance. For example, the extraordinary toughness of bone is due to the combined action of structural elements at the nanometre [34,35] and the micrometre scales [20]. Clearly, hierarchical structuring provides a template for bio-inspired materials synthesis and adaptation of properties for specific functions [36]. Functionally graded materials are examples of materials with a hierarchical structure. New functions may be obtained by structuring a given material, instead of choosing a new material with the desired function. One example for this strategy is composite materials that are omnipresent in nature.

They feature lamellar structures, such as in seashells [25,37,38], glass spicules [32,39] or fibrous structures such as in bone [2,18,20] or wood [2,21–33]. These structures carry many similarities with man-made fibre glass and ceramic laminates. It is highly remarkable that totally different strategies have converged to give similar solutions. Moreover, interfaces play a crucial role in hierarchical composite materials. Joining elements by gluing [35,37,40] is one aspect, while control of the synthesis of components, such as crystals, is another aspect. For a while, this topic has been addressed in the research field of bio-mineralisation [41]. Hierarchical hybrid materials can also provide movement and motility. Muscles and connective tissues are integrated to form complex materials systems, which function as a motor and a supporting structure at the same time. This may inspire material scientists to invent new concepts for active biomimetic materials [42].

Both the biological structure and the set of problems that the structure is capable of solving can inspire us. However, *we may not succeed if we follow without modifications of the solutions found by nature.*

The generation of design (configuration, pattern and geometry) in nature is thought to be a universal physics phenomenon called the constructal law. This law proposed by Bejan and Lorente in 1996 [43] unites the animate with the inanimate over an extremely broad range of scales, from the design of the snowflake, to animal design, and to the tree design of the Amazon river basin.

2.4   Natural self-cleaning and self-healing capabilities

2.4.1   Self-cleaning

The lotus leaf is famous for its self-cleaning behaviour. Raindrops do not pollute their surface but roll off and in this way, they remove powder-like contamination from the leaf surfaces [44,45]. This unique feature of the lotus leaf arises from its low surface energy and its rough surface texture. Inspired by the lotus effect, researchers have successfully fabricated many artificial super-hydrophobic surfaces, defined by a static contact angle above 150° and a roll-off contact angle less than 10°. Here the roll-off angle is defined as the angle at which a water droplet begins to roll down an inclined surface [46]. Self-cleaning surfaces offer many prospective applications in industrial and biological processes [47].

2.4.2   Damage and repair healing

The most intriguing property of the biological materials is probably their capacity to self-heal. At the smallest scale, there is the concept of sacrificial bonds between molecules that break and reform dynamically [38]. Bond breaking and reforming was found, for example, to occur upon deformation of wood [23] and bone [35,48]. This provides the possibility for plastic deformation (without creating permanent damage) as in many metals and alloys. At higher levels, many organisms have the capability to remodel the material. In bone, for example, specialised cells (osteoclasts) are permanently removing material, while other cells (osteoblasts) are depositing new tissue. This cyclic replacement of the bone material has at least two results. First, it allows a continuous structural adaptation to changing external conditions. Second, the damaged material may be removed and replaced by a new tissue [9]. In technical terms, this would mean that a sensor/actuator system is put in place to replace the damaged material wherever needed.

For example, a change in environmental conditions can be partially compensated by adapting the form and microstructure to the new conditions. As a typical example, the growth direction adaptation of a tree after a slight landslide [49,50]. Nature can also heal a fractured or critically damaged tissue. In most cases, wound healing is not a one-to-one replacement of a given tissue, but it rather starts with the formation of an intermediate tissue (based on a response to inflammation), followed by a scar tissue. An exception to this is bone tissue, which is able to regenerate completely and where the intermediate tissue (the callus) is eventually replaced by a material of the original type [51,52]. The science of self-healing materials is still in its infancy [53], but represents a major opportunity for biomimetic materials research.

2.4.3   Biological wound healing in skin

Structural polymers are used in applications ranging from adhesives to coatings to microelectronics to composite airplane wings, but they are highly susceptible to damage in the form of cracks. These cracks are often formed deep within the structure, where detection is difficult and repair is almost impossible. Regardless of the application, once cracks have formed within polymeric materials, the integrity of the structure is significantly compromised.

The addition of self-healing functionality to polymers provides a novel solution to this long-standing problem and represents the first step towards the development of materials systems with greatly extended lifetimes. In biological systems, chemical signals released at the site of fracture initiate a systemic response that transports repair agents to the site of injury and promotes healing. The biological processes that control tissue response to injury and repair are extraordinarily complex, involving inflammation, wound closure and matrix re-modelling [54]. Coagulation and inflammation begin immediately when tissue is wounded. After about 24 hours, cell proliferation and matrix deposition begin to close the wound. During the final stage of healing, the extracellular matrix is synthesised and re-modelled as the tissue regains strength and function. Ideally, synthetic reproduction of the healing process in a material requires an initial rapid response to mitigate further damage, efficient transport of reactive materials to the damage site and structural regeneration to recover full performance.

2.5   Conclusions

Nature provides a wide range of materials with different functions and which may serve as a source of inspiration for the materials scientists. We take the point-of-view that a successful translation of these ideas into the technical world requires more than the observation of nature. A thorough analysis of structure–function relationships in natural tissues must precede the engineering of new bio-inspired materials. There are many opportunities for lessons from the biological world, that is, on growth and functional adaptation, about hierarchical structuring, on damage repair and self-healing. Biomimetic materials research is becoming a rapidly growing and enormously promising field. Serendipitous discovery from the observation of nature will be gradually replaced by a systematic approach involving the study of natural tissues in laboratories, the application of engineering principles to the further development of bio-inspired ideas and the generation of specific databases.

References

[1] D.W. Thompson, *On Growth and Form*, Cambridge University Press, Cambridge, 1968.

[2] J.D. Currey, *Bones: Structure and Mechanics*, Princeton University Press, Princeton, NJ, 2002.

[3] J.F.V. Vincent, *Structural Biomaterials*, Princeton University Press, Princeton, NJ, 1991.

[4] M. Kemp 'Structural intuitions and metamorphic thinking in art, architecture and science', *Metamorph – 9th International Architecture Exhibition Focus*, Fondazione La Biennale di Venezia, Italy, 2004, pp. 30–43.

[5] Ph. Steadman, *The Evolution of Designs, Biological Analogy in Architecture and the Applied Arts*, revised edition, Routledge, London and New York, 2008, pp. 21–53.

[6] W. Nachtigall, *Bionik: Grundlagen und Beispiele für Ingenieure und Naturwissenschaftler*, Springer, Germany, 1998.

[7] G. Jeronimidis and A. Atkins, *Proceedings of the Institution of Mechanical Engineers Part C: Journal of Mechanical Engineering Science*, 1995, **209**, 4, 221.

[8] R. Lakes, *Nature*, 1993, **361**, 6412, 511.

[9] J.D. Currey, *Science*, 2005, **309**, 5732, 253.

[10] P. Fratzl, *Journal of the Royal Society – Interface*, 2007, **4**, 15, 637.

[11] Steven D. Strauss, *The Big Idea: How Business Innovators Get Great Ideas to Market*, Kaplan Business, Dearborn Trade Publishing, Chicago, IL, 2002, pp. 15–18.

[12] Y. Chauvin, R.H. Grubbs, and R.R. Schrock, *Nobel Prize in Chemistry*, 2005. http://www.nobelprize.org/nobel_prizes/chemistry/laureates/2005/press.html

[13] G. Jeronimidis, *Structural Biological Materials, Volume 4: Design and Structure-Property Relationships*, Ed., M. Elices, Pergamon Press, The Netherlands, 2000, p. 19.

[14] G. Nachtrab and K.D. Poss, *Development*, 2012, **139**, 15, 2639.

[15] J.A. Baddour, K. Sousounis and P.A. Tsonis, *Birth Defects Research Part C*, 2012, **96**, 1.

[16] LPX First Lunar Flight of Lunar Plant Growth Experiment, *http:// www.nasa.gov/centers/ames/cct/office/cif/2013/lunar_plant.html*

[17] J-Y. Rho, L. Kuhn-Spearing and P. Zioupos, *Medical Engineering and Physics*, 1998, **20**, 2, 92.

[18] S. Weiner and H.D. Wagner, *Annual Review of Materials Science*, 1998, **28**, 271.

[19] P. Fratzl, H.S. Gupta, E.P. Paschalis and P. Roschger, *Journal of Materials Chemistry*, 2004, **14**, 2115.

[20] H. Peterlik, P. Roschger, K. Klaushofer and P. Fratzl, *Nature Materials*, 2006, **5**, 1, 52.

[21] J. Barnett and G. Jeronimidis, Eds., *Wood Quality and Its Biological Basis*, Blackwell, Oxford, 2003.

[22] B. Hoffmann, B. Chabbert, B. Monties and T. Speck, *Planta*, 2003, **217**, 1, 32.

[23] J. Keckes, I. Burgert, K. Frühmann, *et al.*, *Nature Materials*, 2003, **2**, 12, 810.

[24] M. Milwich, T. Speck, O. Speck, T. Stegmaier and H. Planck, *American Journal of Botany*, 2006, **93**, 10, 1455.

[25] S. Kamat, X. Su, R. Ballarini and A.H. Heuer, *Nature*, 2000, **405**, 6790, 1036.

[26] F. Vollrath and D.P. Knight, *Nature Materials*, 2001, **410**, 6828, 541.

[27] E. Arzt, S. Gorb and R. Spolenak, *Proceedings of the National Academy of Sciences of the United States of America*, 2003, **100**, 19, 10603.

[28] J. Aizenberg, A. Tkachenko, S. Weiner, L. Addadi and G. Hendler, *Nature*, 2001, **412**, 6849, 819.

[29] P. Vukusic and J.R. Sambles, *Nature*, 2003, **424**, 6950, 852.

[30] D. Raabe, C. Sachs and P. Romano, *Acta Materialia*, 2005, **53**, 15, 4281.

[31] D. Raabe, P. Romano, C. Sachs, *et al.*, *Journal of Crystal Growth*, 2005, **283**, 1–2, 1.

[32] J. Aizenberg, J.C. Weaver, M.S. Thanawala, V.C. Sundar, D.E. Morse and P. Fratzl, *Science*, 2005, **309**, 5732, 275.

[33] H. Jinlian, *Structure and Mechanics of Woven Fabrics*, Woodhead Publishing Ltd., Cambridge, 2004.

[34] H. Gao, B. Ji, I.L. Jäger, E. Arzt and P. Fratzl, *Proceedings of the National Academy of Sciences of the United States of America*, 2003, **100**, 10, 5597.

[35] H.S. Gupta, P. Fratzl, M. Kerschnitzki, G. Benecke, W. Wagermaier and H.O.K. Kirchner, *Journal of the Royal Society Interface*, 2007, **4**, 13, 277.

[36] D.A. Tirrell, *Hierarchical Structures in Biology as a Guide for New Materials Technology*, National Academy Press, Washington, DC, 1994.

[37] Z. Tang, N.A. Kotov, S. Magonov and B. Ozturk, *Nature Materials Journal*, 2003, **2**, 6, 413.

[38] G. E. Fantner, E. Oroudjev, G. Schitter, *et al.*, *Biophysical Journal*, 2006, **90**, 4, 1411.

[39] A. Woesz, J.C. Weaver, M. Kazanci, *et al.*, *Journal of Materials Research*, 2006, **21**, 8, 2068.

[40] B.L. Smith, T.E. Schäffer, M. Viani, *et al.*, *Nature*, 1999, **399**, 6738, 761.

[41] S. Mann, *Biomineralization: Principles and Concepts in Bioinorganic Chemistry*, Oxford University Press, Oxford, 2001.

[42] A. Sidorenko, T. Krupenkin, A. Taylor, P. Fratzl and J. Aizenberg, *Science*, 2007, **315**, 5811, 487.

[43] A. Bejan, and S. Lorente, *Journal of Applied Physics*, 2013, **113**, 15, 151301.

[44] W. Barthlott and C. Neinhuis, *Planta*, 1997, **202**, 1, 1.

[45] L. Gao and T.J. McCarthy, *Langmuir*, 2006, **22**, 7, 2966.

[46] M. Nosonovsky and B. Bhushan, *Nano Letters*, 2007, **7**, 9, 2633.

[47] B.D. Hatton and J. Aizenberg, *Nano Letters*, 2012, **12**, 9, 4551.

[48] H.Y. Erbil, A.L. Demirel, Y. Avci and O. Mert, *Science*, 2003, **299**, 5611, 1377.

[49] J.B. Thompson, J.H. Kindt, B. Drake, H.G. Hansma, D.E. Morse and P.K. Hansma, *Nature*, 2001, **414**, 6865, 773.

[50] C. Mattheck and K. Bethge, *Naturwissenschaften*, 1998, **85**, 1, 1.

[51] C. Mattheck and H. Kubler, *Wood – The Internal Optimization of Trees*, Springer, Berlin, 1995.

[52] D.R. Carter and G.R. Beaupré, *Skeletal Function and Form: Mechanobiology of Skeletal Development, Aging, and Regeneration*, Cambridge University Press, Cambridge, 2001.

[53] S.R. White, N.R. Sottos, P.H. Geubelle, *et al.*, *Nature*, 2001, **409**, 6822, 794.

[54] A.J. Singer and R.A.F. Clark, *New England Journal of Medicine*, 1999, **341**, 10, 738.

Chapter 3
Theoretical models of healing mechanisms

Modelling of the nature is developed at three levels. At the first level, a relatively simple approach is used to imitate a natural function such as healing human skin, which can then be used for healing a crack.

At the second level, various models are created to produce a multifunctional component, for example, a biomimetic of shark skin, which can be adopted for a swimming suit that gives an increase in the swimming speed. At the same time the textile would be capable of repairing itself after a scratch or a puncture. At the third level, a model will be based on a more complex design.

Most of the models developed try to predict and optimise self-healing behaviour of materials at the first level. Some of the most interesting work in that area is reviewed in Sections 3.1 and 3.2 and gives an example of such modelling with finite element analysis (FEA). No work has been reported on the second level models.

Recently several approaches at third level modelling have been proposed and developed. A brief summary of these models is reviewed in Section 3.3.

3.1   The first level models

Several works on theoretical modelling and utilisation of computational tools to predict and optimise self-healing behaviour of materials are found in the literature.

In one of the earlier investigations, Barbero *et al.* [1] proposed a model based on a continuum thermodynamic framework to predict the self-healing processes. The model was based on a set of equations obtained through the thermodynamic approach, which involved such variables as the damage, plasticity and healing mechanism. These relationships and the damage evolution equations were defined by a nonlinear differential problem and were solved by the means of numerical algorithms. The results allowed validation of the numerical model for a fibre reinforced polymer matrix composite that was subjected to in-plane shear loading. The model accurately predicted the experimental data for samples that did not include a healing agent. The model with a healing agent could not be validated because of the lack of experimental data.

Modelling of fatigue cracks in self-healing polymers at the molecular scale was carried out by Maiti *et al.* [2]. The cure of the healing agent was modelled using a coarse-grain molecular dynamics process. The incorporation of healing kinetics (i.e., degree and rate of cure) in the model enabled the competition between the fatigue crack growth and crack healing mechanisms to be studied. This study of the effect of different loading and healing parameters showed a good qualitative agreement between the experimental observations and the simulation results [2].

Privman *et al.* [3] employed a continuum rate equation modelling to understand self-healing of composites, which were reinforced with nanoporous microglass fibres. The fibres, with an aspect ratio around 3, contained a healing liquid to reduce the fatigue damage. In their earlier work [4], they successfully synthesised nanoporous micron-sized glass capsules. The capsule size was 2 μm in diameter and 5 μm in length, with pores of uniform diameter of 3 nm. In their modelling, the authors focused on the gradual formation of microcracks due to fatigue as well as the healing by nanoporous fibre rupture and release of healing agent. Their rate equation modelling suggests that one of the important challenges will be the size of capsules with healing agents. Larger capsules may fill-in larger cracks. However, the larger capsules may affect the overall strength of the material [3]. This suggestion is consistent with the available experimental observations that a significant amount of microcapsules embedded in the material may actually weaken its mechanical properties.

Zhou *et al.* [5] carried out computational simulations to investigate the effect of embedded microvascular vessels (or hollow spheres) on the mechanical properties of self-healing composites. Their work was supported by experimental observations. Referring to the experimental investigation of Kousourakis [6], the authors noticed that the incorporated microvascular hollow glass fibres (of varying diameter up to 680 μm) into the cross-ply carbon fibre composites had no significant influence on the tension and the compression modulus. However, the strength properties were reduced due to the ply waviness and found to be dependent on the diameter and orientation of the fibres. On the other hand, the interlaminar fracture toughness was improved up to 50% even without triggering self-healing. In their finite element modelling, the authors [5] embedded a 680 μm diameter hollow fibre. The stress concentration around the hollow fibre and the damage progression was determined using the Abaqus finite element model. Their computations showed that the presence of the hollow fibre increased the tensile stress up to 24% (in the 0° ply). The findings of their computations can be useful for microcapsule-based self-healing of polymer composites.

Mookhoek *et al.* [7] studied the effect of liquid-based healing systems of aspect ratio, volume fraction and orientation of elongated capsules in comparison to the spherical capsules on the healing efficiency. Their simulation results showed that, for spherical capsules, the amount of healing agent released per unit crack area decreases rapidly with decreasing capsule size. They recommended that small, for example, spherical nano-capsules, should only be used for healing of small scale damage (such as interfacial debonding) but not for enhancing the healing efficiency [7]. Their model, on the other hand, predicted significantly more release of healing agent for the elongated capsules. They also observed that the pre-ferred orientation of elongated capsules perpendicular to the fracture plane, instead of random orientation, offers the possibility of the lower capsule loading for the specified healing efficiency. Moreover, the host matrix properties were found to be less influenced by the elongated capsules than the spherical capsules. Contrary to their observations, the recent work of Lv and Chen [8] revealed that the hitting probability of elongated capsules with a higher aspect ratio is not always larger than that for spherical cap-sules. One of the interesting results of Mookhoek *et al.*'s [7] work is that they found the crack width, instead of the crack length, to be the key factor in the healing process. There is an optimal crack width that can be healed efficiently. Moreover, the maximum crack width that can be subjected to healing is only dependent on the material properties of the liquid/matrix system and not on the capsule size, volume fraction or geometry. The critical width is given by (3.1) [7]:

$$\mathrm{PCOD} = 2((2\gamma_1(1 - \cos\beta))/g\rho)^{0.5} \tag{3.1}$$

where PCOD is the parallel crack opening distance, γ_1 is the surface tension of liquid, β is the liquid-surface contact angle, g is the gravitational constant and ρ is the liquid density.

As an example, the critical crack width that can be healed for the epoxy-dicyclopentadiene system ($\gamma_1 = 0.036$ J/m^2, $\rho = 987$ kg/m^3, $\beta = 4.5°$) was calculated to be around 340 µm [7]. This finding could be of great importance for research on a liquid-based self-healing system.

Lv *et al.* [9] proposed a two-dimensional (2D) analytical model for computing the probability that a regular crack pattern hits the capsules embedded in cementitious materials. They used the concept of integral geometry and probability distribution and presented a theoretical solution of the exact dosage of capsules. These materials were simplified as linear cracks and zonal cracks in the 2D plane. The models were validated using computer simulations. In their subsequent works [10,11], the same authors extended the models for various three-dimensional (3D) crack patterns. However, the problem was not theoretically solved for the random distribution of cracks in the matrix, which is usually the case for polymer composite materials.

In a similar work, Zemskov *et al.* [12] presented two 2D analytical models for computing the probability that a crack hits an encapsulated particle. The models relate the size of the capsules, content ratio (volume fraction), average intercapsule distance and the depth of crack to the probability of hitting the capsules. The study was performed on the basis of self-healing of concrete with embedded macrocapsules containing a bacterial-based healing agent.

In one of the models, the spherical macrocapsules were placed in a sequence of layers in the bulk materials (Model I). In the other model (Model II), the capsules were placed in a fully randomised fashion. Elementary concepts of probability and planar geometry were applied to evaluate the hitting probabilities for Model I, while advanced statistical probability concepts involving rigorous mathematical computations were applied in Model II. According to Lv and Chen [8], the hitting probability formula proposed by Zemskov *et al.* [12] for Model II is just an abstract expression and cannot be analytically verified. However, in Zemskov *et al.*'s work [12], both models were validated with Monte Carlo simulations. Model II predicted a lower density of capsules for the same probability of hitting as compared to Model I. Lv and Chen [8] noticed that in their Model II, the ratio of the size of capsule to the size of cubic sample is not large enough, which may reduce the randomness of capsules. In other words, the randomness of distribution of capsules in the self-healing model would not be guaranteed [8]. One of the findings of these models is that the capsules with lower radii require a lower content ratio (volume fraction) for the specified probability and depth of crack. The models, however, did not consider the physical characteristics of the capsules and the matrix [12].

Lv and Chen [8] in their most recent work developed both 2D and 3D models based on the concept of integral geometry and geometric probability. They used the models to determine the probability of hitting and dosage of capsules (randomly distributed in the matrix) required to repair discrete cracks that occur independently in matrix. A probabilistic healing approach was proposed which would provide the required volume fraction of capsules for a targeted level of healing. This model provides explicit expressions for the hitting probabilities and dosage of capsules in terms of the sizes of cracks and capsules. The authors also investigated the effect of different aspect ratios of the elongated capsules on the hitting probabilities. Their models were validated by Monte Carlo simulations, which showed a good agreement between the model predictions and computer simulations. For small cracks the hitting probability of elongated capsules is not always larger than that of spherical capsules. Experimental data, however, are required to confirm the findings.

Verberg *et al.* [13] modelled a fluid driven microcapsule to determine how the release of the encapsulated nanoparticles could be harnessed to repair damage on the underlying surface. The simulations revealed that these microcapsules can deliver the encased materials to specific sites on the substrate. Once the healing nanoparticles were deposited on the desired sites, the fluid-driven capsules could move further along the surface. In their subsequent work [14], they extended the 2D model to simulate the 3D interactions of deformable microcapsules with a substrate that contains a 3D crack. The ability to fabricate amphiphilic capsules that encapsulate nano-particles inspired the authors to formulate a delivery system that could be transported by a flow in an aqueous solution and specifically target the nanoparticles to the hydrophobic domains [14]. The work can be used as a guideline for designing particle-filled microcapsules that can be used to repair in a continual fashion. In addition, such an approach may be used to repair the damage in microchannels and microfluidic devices where a continuous fluid motion is maintained. Some other examples of computational studies in self-healing materials can be found in [15–18] (Table 3.1).

3.2 Example of modelling with finite element analysis (ANSYS code)

As an example of the first level model, a study of the self-healing of resin containing embedded microcapsules filled with a healing agent, and submitted to a crack is presented. The prepared samples have the shape of a tapered double cantilever beam (TDCB), commonly used to measure the mechanical properties of materials (Figure 3.1(a)) The TDCB have standard dimensions (more details are given in Chapter 5). The mechanical properties of the samples are determined by pulling the sample in two vertically opposite directions using the two holes, while simultaneously measuring the applied force (load) and the displacement. Commonly a material testing system (MTS) machine is used for these tests. During the test, the applied load increases, the strain follows, passing from elastic regime to inelastic regime, then to irreversible damage (crack). TDCB was modelled by

Table 3.1   Summary of the proposed models and their validation

Model and reference	Validation
Thermodynamic framework with variables representing the damage, plasticity and healing mechanism (Barbero *et al.* [1])	Fibre reinforced polymer matrix composite subjected to in-plane shear loading without healing agents
Molecular-level modelling of fatigue crack and competition between fatigue crack growth and crack healing (Maiti *et al.* [2])	Good qualitative agreement between experimental observations and simulation
Curing rate equation modelling (Privman *et al.* [3])	Composites reinforced with nanoporous microglass fibres (concluded it is better to use smaller microcapsules)
FEA model (Abaqus) embedded microvascular vessels (Zhou *et al.* [5])	No significant influence on the tension and compression modulus
Elongated capsules to heal larger cracks (Mookhoek *et al.* [7])	Validated by Monte Carlo simulations
2D model based on integral geometry and probability distribution (Lv *et al.* [9])	Validated by Monte Carlo simulations
2D model relates the size of capsules, content ratio (volume fraction), average inter-capsule distance and the depth of crack to the probability of hitting of the capsules (Zemskov *et al.* [12])	Validated with Monte Carlo simulations
Model based on a fluid-driven, particle-filled microcapsule along adhesive substrate (Verberg *et al.* [13])	Model under development

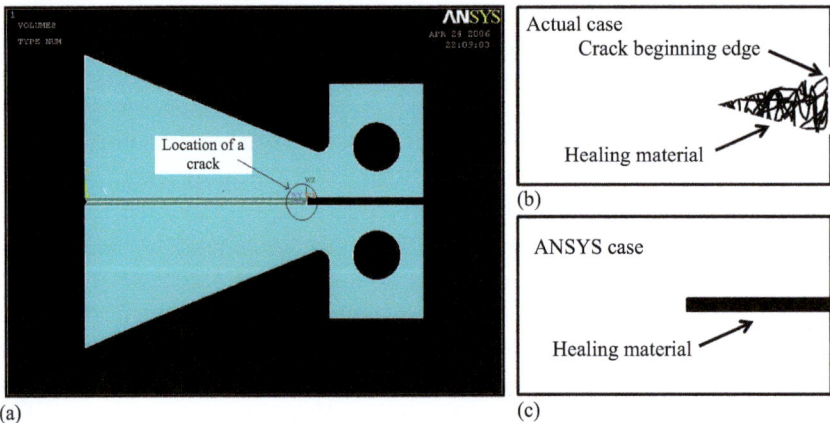

(a)   (c)

Figure 3.1   (a) TDCB specimen as modelled in ANSYS, the encircled portion showing the crack location; (b) enlarged view of this portion with 100% healed crack in a real case and (c) a healed model in ANSYS

ANSYS, which is a known engineering simulation software based on finite element analysis, offering solution sets in engineering simulation that a design process requires.

To measure the healing efficiency of our samples, a cut of about 2 cm was made with a blade in the middle of the TDCB (Figure 3.1) to simulate a crack. Then the samples with/without healing agent were tested with an MTS machine and their mechanical properties were compared.

The stiffness parameter was considered as the representative parameter to evaluate the healing efficiency.

The simulated stiffness was deduced from the FEA code (ANSYS):

- A crack of desired length was simulated in the TDCB model (Figure 3.1). Table 3.2 gives the value of crack lengths used.
- A portion of the crack was filled to represent the healed crack (Figure 3.2(a)). The TDCB portion was meshed and indicated by zone

Table 3.2 Parameters used in ANSYS analysis for the healing modelling of the crack

Element parameter in ANSYS code	Value
E_1: Young's modulus of TDCB sample (resin 828)	3.1, 3.3 and 3.5 GPa
v_1: Poisson's ratio of TDCB sample (resin 828)	0.2
E_2: Young's modulus of healed material	1, 0.1 and 0.01 GPa
v_2: Poisson's ratio of healed material	0.3
Crack length	20, 22, 24, 26 and 28 mm
Healed efficiency (or fraction)	0%, 25%, 50%, 75% and 100%

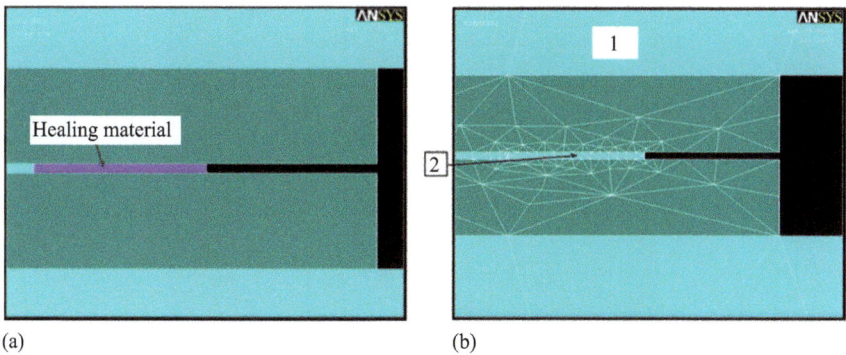

(a)　　　　　　　　　　　　　　　　　(b)

Figure 3.2　(a) Sample with crack length a = 22 mm and 50% healed crack and (b) sample showing a different numbering system for TDCB sample material and healing material

'1'. The healing material was meshed and is indicated by zone '2' (Figure 3.2(b)). Table 3.2 summarises the corresponding Young's modulus and Poisson's ratio values used in the model.

- A load was applied to simulate the MTS machine. The upper arm of the TDCB was fully constrained, and a tensile load was applied on the lower arm (Figure 3.3(a)). The model provided the displacement (Figure 3.3(b)), perpendicular to the crack plane.
- ANSYS analysis was performed for 0%, 25%, 50%, 75% and 100% healing efficiency. The 0% healing represented the original sample without a healing agent.

An example of the simulation results is presented in Figure 3.4. The data obtained by the iterative simulation of different values are given in

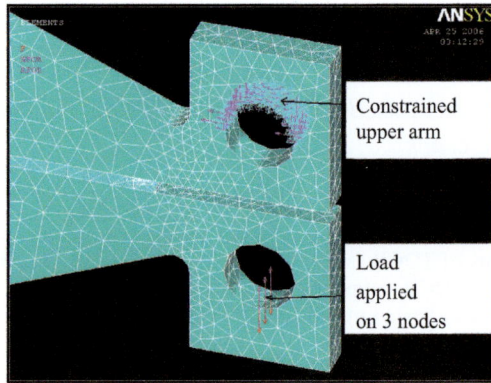

(a)

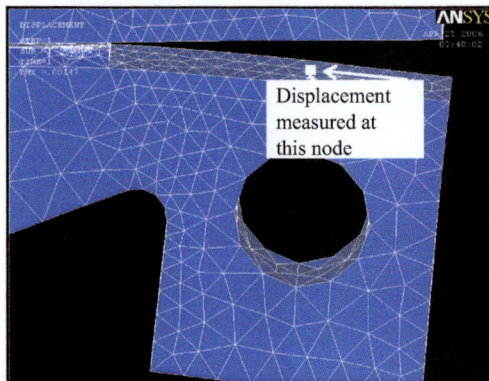

(b)

Figure 3.3   *(a) Constrained and loaded TDCB model and (b) sample showing the location where of node displacement perpendicular to the crack plane were measured (in the y direction)*

(a)

(b)

(c)

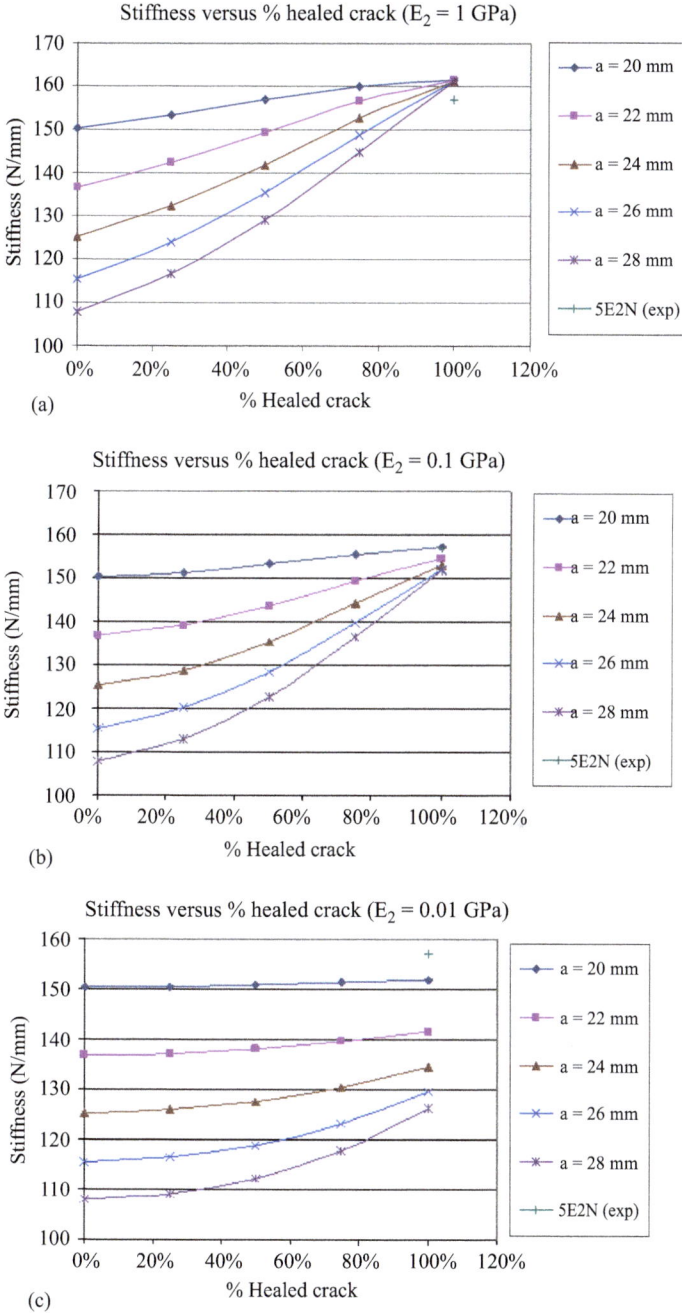

Figure 3.4   *Stiffness as a function of the healed crack percentage (%)*
          graphs for different E_2 values: (a) 1 GPa, (b) 0.1 GPa and
          (c) 0.01 GPa ($E_1 = 3.5$ GPa) compared with the experimental
          measurement

Table 3.2. The optimal set of stiffness values is the one which is the closest possible to the experimentally measured parameters (Figure 3.4(b)). Young's modulus of the healed materials $E_2 = 0.1$ GPa.

3.3   Third level models

Nature offers complex multifunctional systems based on very simple designs. One can give examples of Velcro sticking, a gecko moving on a vertical wall and the self-cleaning of a lotus leave. This raises the question of how to find more correlation between engineering and biological systems in order to take advantage of the natural designs.

The correlation method was originally developed to find similarities between two engineering systems, and take the advantages of the existing models and theories for one system and adapt them to the other. This approach is applied to learn simplicity, multifunctionality and adaptability (self-healing) from natural designs. The challenges are to find simple engineering designs that could perform similar multifunction. The most common approaches require going through a systematic methodology and to compare the processes designed in engineering and in nature and their functional structures. It was noticed that there were large differences in the functional terms used in engineering and in biology. The idea of a standard set of engineering functional terms for creating functional structures was originally proposed by Pahl *et al.* [19]. In their proposal a function would represent an operation performed on a flow of material, signal or energy. The idea was gradually accepted. In 2002, Hacco and Shu [20] developed a method to search biomimetic conceptual design to find solutions for a specific manufacturing problem. Their method was refined by Chiu and Shu who looked for design inspiration by searching the biological literature using functional keywords [21]. Nagel *et al.* [22] developed the model further by creating *An Engineering-To-Biology Thesaurus*. The objective is to assist engineers by correlating biological terms to engineering terms. The thesaurus proposed eight classes of functions and three classes of flows, each class having an increase in the specification at the secondary and tertiary levels, with a corresponding list of biological functions (Tables 3.3–3.5). Cheong *et al.* [23] further developed the previous model with an algorithm that identifies keywords, which are more effective in searching biological texts.

The codes performing a systematic search of analogies between nature and engineering are continuously improved. Table 3.3 presents the list of the function and flow classes with their secondary levels. Table 3.4 provides the number of secondary and tertiary functions and flows. Table 3.5 gives

Table 3.3   List of the function and flow classes with their secondary levels

Class	Secondary level
	Function
Branch	Separate and distribute
Channel	Import, export, transfer and guide
Connect	Couple and mix
Control magnitude	Actuate, regulate, change and sop
Convert	Convert
Provision	Store and supply
Signal	Sense, indicate and process
Support	Stabilise and secure
	Flow
Material	Human, gas, liquid, solid and mixture
Signal	Status and control
Energy	Human acoustic, chemical, electrical, electromagnetic, hydraulic, magnetic, mechanical and pneumatic thermal

Table 3.4   Number of functions and flows

Class	Functions	Flows
Primary classes	8	3
Secondary classes	21	20
Tertiary classes	24	22

Table 3.5   Example of the class of energy with secondary and tertiary level

Class	Secondary	Tertiary	Corresponding biological function
Energy	Human	–	Being and body
	Acoustic	–	Echolocation and sound wave
	Chemical	–	Calorie, metabolism, glucose, glycogen, ligand, nutrient, starch, fuel, sugar, mitochondria, lipid and gibberellin
	Electrical	–	Electron, potential, feedback, charge and field
	Electromagnetic	Optical	Light and infrared
		Solar	Light, sun and ultraviolet light
	Hydraulic	–	Pressure, osmosis and osmoregulation
	Magnetic	–	Gravity, field and wave
	Mechanical	–	Muscle contraction, pressure, tension, stretch and depress
		Rotational	Protein folding
		Translational	–
	Pneumatic	–	Pressure
	Thermal	–	Temperature, heat, infrared and cold

an example of energy function with its secondary and tertiary levels, with the list of corresponding biological functions.

References

[1]   E.J. Barbero, F. Greco and P. Lonetti, *International Journal of Damage Mechanics*, 2005, **14**, 1, 51.

[2]   S. Maiti, C. Shankar, P. Geubelle and J. Kieffer, *Journal of Engineering Materials and Technology*, 2006, **128**, 4, 595.

[3]   V. Privman, A. Dementsov and I. Sokolov, *Journal of Computational and Theoretical Nanoscience*, 2007, **4**, 1, 190.

[4]   Y. Kievsky and I. Sokolov, *IEEE Transactions on Nanotechnology*, 2005, **4**, 5, 490.

[5]   F. Zhou, C. Wang and A. Mouritz, *Materials Science Forum*, 2010, **654**, 2576.

[6]   A. Kousourakis, *Mechanical Properties and Damage Tolerance of Aerospace Composite Materials Containing CVM Sensors*, PhD Thesis, RMIT University, Melbourne, Australia, 2008.

[7]   S.D. Mookhoek, H.R. Fischer and S. Van der Zwaag, *Computational Materials Science*, 2009, **47**, 2, 506.

[8]   Z. Lv and H. Chen, *Computational Materials Science*, 2013, **68**, 81.

[9]   Z. Lv, H. Chen and H. Yuan, *Science and Engineering of Composite Materials*, 2011, **18**, 1–2, 13.

[10] Z. Lv, H. Chen and H. Yuan, *Materials and Structures*, 2011, **44**, 5, 987.

[11] Z. Lv, H. Chen and H. Yuan, *Journal of Intelligent Material Systems and Structures*, 2014, **25**, 1, 47.

[12] S.V. Zemskov, H.M. Jonkers and F.J. Vermolen, *Computational Materials Science*, 2011, **50**, 12, 3323.

[13] R. Verberg, A.T. Dale, P. Kumar, A. Alexeev and A.C. Balazs, *Journal of Royal Society – Interface*, 2007, **4**, 13, 349.

[14] G.V. Kolmakov, R. Revanur, R. Tangirala, *et al.*, *ACS Nano*, 2010, **4**, 2, 1115.

[15] O. Herbst and S. Luding, *International Journal of Fracture*, 2008, **154**, 1–2, 87.

[16] K.A. Smith, S. Tyagi and A.C. Balazs, *Macromolecules*, 2005, **38**, 24, 10138.

[17] D.S. Burton, X. Gao and L.C. Brinson, *Mechanics of Materials*, 2006, **38**, 5–6, 525.

[18] G. Thatte, S.V. Hoa, P.G. Merle, E. Haddad and Y. Guntzberger, *Proceedings of the First International Conference on Self-healing*

Materials, The Delft Centre for Materials, Delft University of Technology, Noordwijk, The Netherlands, 2007.

[19] G. Pahl, W. Beitz, J. Feldhusen and K-H. Grote, *Engineering Design: A Systematic Approach*, 3rd edition, Springer-Verlag, London, 2007.

[20] E. Hacco and L.H. Shu, *Proceedings of the ASME Design Engineering Technical Conference*, Volume **3**, Montreal, Canada, 2002, p. 239.

[21] I. Chiu and L.H. Shu, *Artificial Intelligence for Engineering Design, Analysis and Manufacturing*, 2007, **21**, 1, 45.

[22] J.K.S. Nagel, R.B. Stone and D.A. McAdams, *Proceedings of the ASME International Design Engineering Technical Conferences*, Volume **5**, Montreal, Quebec, Canada, 2010, p. 117.

[23] H. Cheong, R.B. Stone, D.A. McAdams, I. Chiu and L.H. Shu, *Journal of Mechanical Design*, 2011, **133**, 2, 021007.

Chapter 4

Self-healing of polymers and composites

Since the first report of the self-repairing composites systems in the literature [1], a conventional strategy was developed by embedding a microencapsulated liquid healing agent and solid catalytic chemical materials within a polymer matrix. Thus, when there is damage induced cracking in the matrix, the microcapsules release their encapsulated liquid healing agent into the crack planes. All the materials involved must be carefully engineered. For example, the encapsulation procedure must be chemically compatible with the reactive healing agent, and the liquid healing agent must not diffuse out of the capsule shell during its shelf-life. At the same time, the microcapsule walls must be resistant to the processing conditions of the host composite. At the same time excellent adhesion with the cured polymer matrix has to be maintained to ensure that the capsules rupture upon composite fracture.

4.1 Microcapsules

Polymeric microcapsules are often prepared *via* a mini-emulsion polymerisation technique, as described by Asua [2]. The procedure involves the well-known oil-in-water dispersion mechanism of the polymeric material. In the majority of self-healing composite systems that have been studied, the microcapsules are made by a urea-formaldehyde (UF) polymer, encapsulating dicyclopentadiene (DCPD) as the liquid healing agent [1,3–9] and/or an epoxy resin [10–14]. For DCPD, during the *in situ* polymerisation process, urea and formaldehyde react in the water phase to form a low molecular weight prepolymer; when the weight of this prepolymer increases, it deposits at the DCPD-water interface. This UF polymer becomes highly crosslinked and forms the microcapsule shell wall. The UF prepolymer particles are then deposited on the surface of the microcapsules, providing a rough surface morphology that aids in the adhesion of the microcapsules with the polymer matrix during composite processing [15]. Moreover, composites using DCPD-filled UF microcapsules have been shown to have a healing ability in monotonic fracture and fatigue [1,3–8].

4.1.1   *Effects of the size and the materials of microcapsules on self-healing reaction performance*

In 2003, Brown *et al.* [9] reported that the microcapsules made in this *in situ* process have an average size of 10–1,000 μm in diameter, with a smooth inner shell thickness of 160–220 nm, and a fill content up to 83%–92% of the liquid healing agent. The mechanical rupture of the microcapsule is the essential condition for the healing process. Thus, it is important to fabricate microcapsules with optimal mechanical properties and wall thickness. The relationship between the stiffness of the capsule and the polymer matrix determines how the crack will propagate in the sample. Keller and Sottos [16] have described how a capsule, which has higher elastic modulus than that of the polymer matrix material, should create a stress field that tends to deflect cracks away from the capsule. A more compliant shell wall, on the other hand, will produce a stress field that attracts the crack towards the microcapsule.

The influence of microcapsule diameter and crack size on the perfor-mance of self-healing materials was investigated by Rule *et al.* [17]. They used an epoxy-based material containing embedded Grubbs' catalyst par-ticles and microencapsulated DCPD. The amount of liquid that micro-capsules could deliver to a crack face was shown to scale linearly with the microcapsule diameter (and thus, the volume) for a given weight of cap-sules. The size of the microcapsule also plays a role in the performance of the system, in terms of the effect on the toughness of the composite and the nature of the interface between the microcapsule and the polymer matrix. Based on these relationships, the size and weight fraction of the micro-capsules can be chosen appropriately to give optimal healing reaction of a predetermined crack size. As mentioned earlier, the wall thickness of the microcapsule is another critical parameter. If the shell wall is too thick, the microcapsule will not easily rupture and healing will not occur. If the shell wall is too thin, however, the microcapsules can rupture during composite manufacture and processing, or the healing agent could leak or diffuse into the matrix. As noted by Brown *et al.* [9], the shell wall thickness is largely independent of the manufacturing parameters and is typically between 160 and 220 nm thick; however, slight adjustments can be made during the encapsulation procedure to alter the resulting microcapsules. The micro-capsule size is controlled mainly *via* the rate of agitation during the encapsulation process. Typical agitation rates reported by Brown *et al.* [9] range from 200 to 2,000 rpm, with finer emulsions, and therefore, smaller diameter capsules are produced with increasing agitating rates. Brown *et al.* [5] noted that smaller microcapsules exhibit maximum toughening at lower concentrations. On the other hand, Rule *et al.* [17] reported that specimens, which contain larger microcapsules perform better than those with smaller microcapsules at the same weight fraction, presumably due to the amount of healing agent present in the specimen. In the latter study, the best healing achieved was on a specimen containing 10 wt% of 386 μm diameter capsules, which corresponds to 4.5 mg of healing agent being delivered per unit crack area (assuming all capsules in the crack plane rupture). The amount of healing agent available for delivery to the crack plane was calculated based on the microcapsule size and weight fraction incorporated into the composite and verified by comparing the data from these autonomously healing samples with that of samples in which a known volume of healing agent. The healing agent was manually injected into the crack plane to initiate the healing process.

With the objective to synthesise smaller microcapsules that exhibit maximum toughening at lower concentrations, Blaiszik *et al.* [18] developed an *in situ* encapsulation method, demonstrating over an order of magnitude size reduction of UF capsules. The capsules with diameters down to 220 nm filled with a DCPD healing agent were successfully produced using sonication techniques and an ultrahydrophobe to stabilise the DCPD droplets. The capsules possess a uniform UF shell wall (77 nm average thicknesses) and display good thermal stability. However, there are several drawbacks to using UF microcapsules. First, the formation of agglomerated nanoparticle debris, which could act as crack initiation sites within the host matrix. Second, rough and porous wall surfaces formed by agglomerated nanoparticles, which may reduce the adhesion between the microcapsules and matrix. Finally, rubbery and thin capsule walls (160–220 nm [9]), which lead to the loss of core material during storage and cause handling difficulties during the processing of the composites. The formation of rough porous capsule surfaces is known to be a common feature in the production UF microcapsules [19]. In addition to the UF microcapsules, melamine-formaldehyde [20,21] and polyurethane (PU) [22] shell wall materials were used to prepare microcapsules of various healing materials. Liu *et al.* [23] have produced microcapsules for self-healing applications with a melamine urea formaldehyde (MUF) polymer shell containing two different healing agent candidates. These agents were 5-ethylidene-2-norbornene (ENB) and ENB with 10 wt% of a norbornene-based crosslinking agent obtained by *in situ* polymerisation in an oil-in-water emulsion. The microcapsules were found to be thermally stable up to 300 °C and exhibited a 10%–15% weight loss when held isothermally at 150 °C for 2 hours. Overall, these MUF microcapsules exhibited superior properties when compared to the UF microcapsules used extensively for self-healing composites. Their manufacturing process is simpler than that for those made from UF. In addition, the microcapsules for such applications should have optimum mechanical strength and rupture under a given mechanical load. Theoretically, the mechanical strength of microcapsules is determined by their chemical composition, structure, size and shell thickness. It has been widely reported that DCPD can be encapsulated using UF. The performance of the self-healing microcapsules depended on their size [17,24]. However, little work has been done on encapsulation of DCPD using different shell materials and on direct measurement of the mechanical strength of single microcapsules containing a self-healing agent [16]. The mechanical strength of microcapsules made of different shell materials, including melamine for-maldehyde (MF) and UF with different core materials, have been characterised by a micromanipulation technique [25–27]. It has been found that MF microcapsules are stronger and ruptured at larger deformations than UF for a given size and shell thickness [27]. It is likely that the strength and deformability of microcapsules depend on their formulation and processing conditions. Therefore, MF microcapsules provide a bigger scope for for-mulation and processing conditions than UF to produce optimum mechanical properties, which may be desirable for self-healing applications.

It is well known that one of the major difficulties in microcapsule manu-facture is to obtain capsules small enough for many practical applications and with a size distribution narrow enough to control release [28]. Various appli-cations have been attempted with more or less success. Microcapsules have been used in the paper industry for a range of different purposes [29], for example, in self-copying carbonless copy paper [30], and in the food and packaging industries for applications such as control of aroma release and as temperature or humidity indicators [31,32]. Other possible applications might include encapsulation of antimicrobial agents or scavengers in active packa-ging. Recently, Andersson *et al.* [33] have developed microcapsules with a hydrophobic core surrounded by a hydrophobically modified polysaccharide membrane in aqueous suspension. The idea was to obtain capsules fulfilling both the criteria of small capsule size and reasonably high solids content.

The innovative work of Mookhoek *et al.* [34] in which microcapsules with a PU shell and a size of around 1.4 μm are filled with dibutylphthalate (DBP) and used as Pickering stabilisers to create larger (~140 μm) micro-capsules containing a second liquid phase of DCPD. The binary micro-capsules were made by encapsulating the dispersed DCPD liquid (stabilised with the UF containing DBP microcapsules in water) *via* an isocyanate–alcohol interfacial polymerisation reaction.

Recently, very small microcapsules ENB with a diameter of about 1 μm and a thin shell of ENB/poly-MUF (PMUF) were obtained [35]. The char-acteristics of these small microcapsules of about 1–2 μm in diameter are shown in Figures 4.1–4.6.

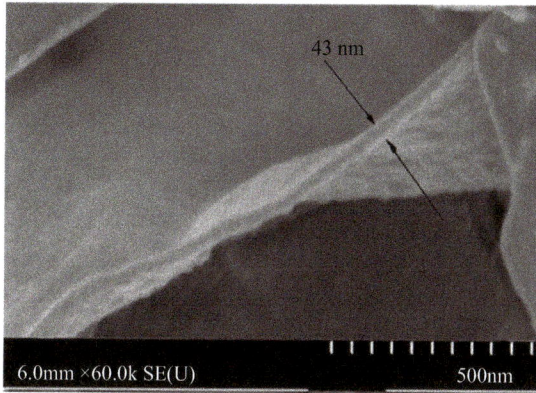

Figure 4.1 Typical scanning electron microscopy micrograph of the PMUF shell of 1 µm diameter capsules. The shell thickness is about 43 nm

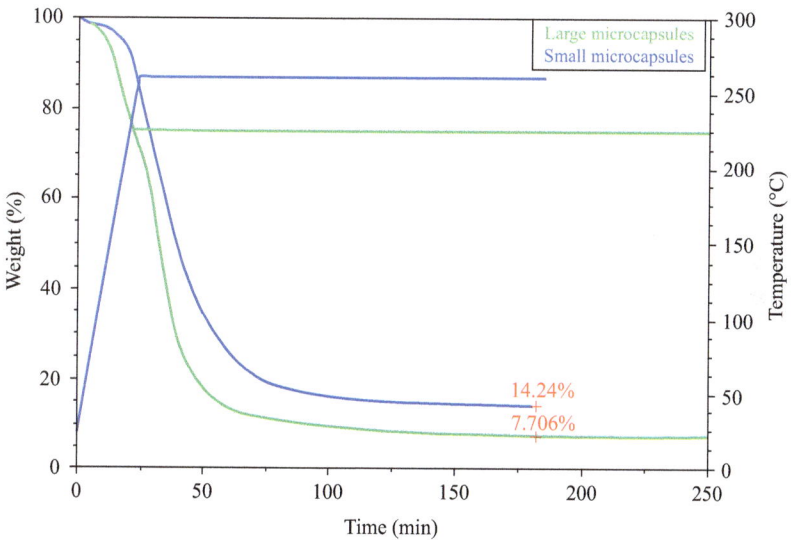

Figure 4.2 Weight loss of the self-healing agent in PMUF microcapsules measured by thermogravimetry analysis

4.1.2 Retardation of fatigue cracks

To slow down the growth of fatigue cracks, Keller *et al.* [36] have shown that the incorporation of self-healing functionality in a polysiloxane elastomer successfully reduced, by 24%, the growth of fatigue cracks under

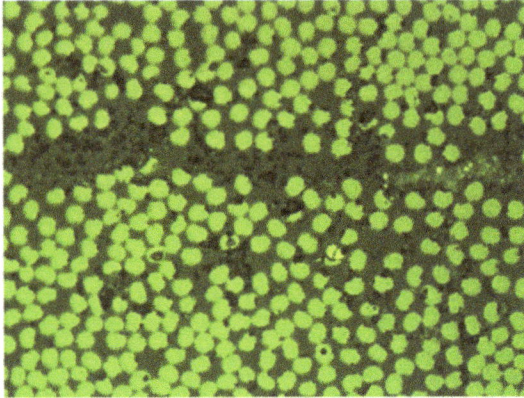

*Figure 4.3   Cross-section of an optical view of two layers of CFRP
laminates containing 15 wt% of microcapsules*

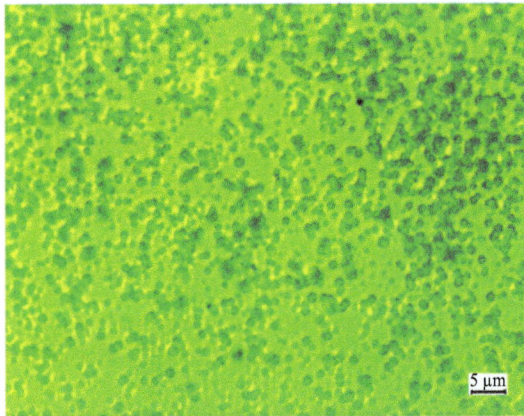

*Figure 4.4   Optical image of typical small microcapsules prepared using
ultrasound for 20 minutes. Image taken at 1,600 magnification*

torsion fatigue loading. This retardation was attributed, in part, to a sliding-
crack-closure mechanism, where polymerised healing agent shields the
crack tip from the applied far-field stress. On the other hand, shape-memory
alloy (SMA) wires are well-suited to this application, since they exhibit a
thermoelastic martensitic phase transformation, contracting above their
transformation temperature and exerting large recovery stresses of up to 800
MPa when constrained at both ends [37,38]. Moreover, Rogers *et al.* have
shown that, when an SMA wire is embedded within an epoxy matrix, the
full recovery force acts at the free edges of the component [39]. Therefore,

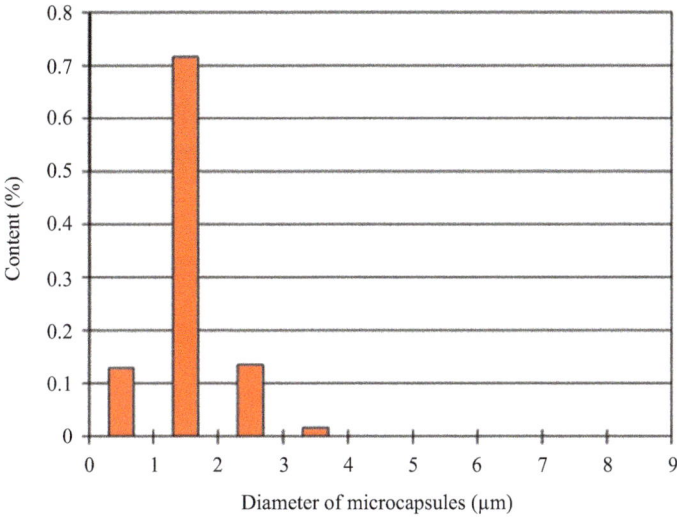

Figure 4.5   Distribution histogram of the ENB monomer encapsulated into PMUF shells prepared using ultrasonic at L-1.5 for 20 min at 1600 magnification

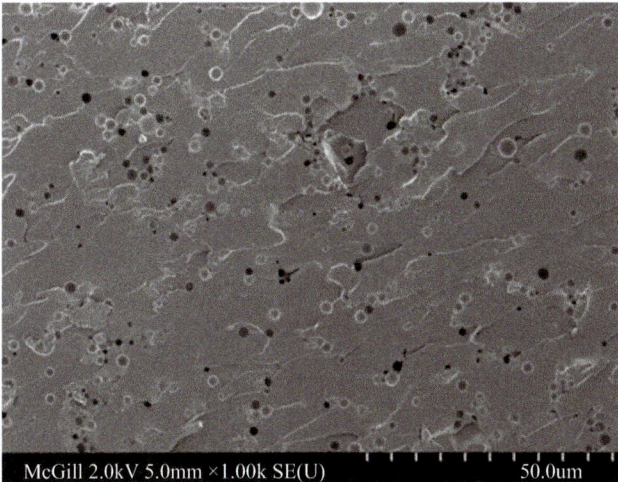

McGill 2.0kV 5.0mm ×1.00k SE(U)   50.0um

Figure 4.6   Typical SEM image of 10 wt% of ENB/PMUF microcapsules within the Epon^{TM} 828 epoxy resin

an SMA wire bridging a crack should induce a large closure force. Indeed, Kirkby *et al.* [40] have reported on self-healing polymers with embedded SMA wires, where the addition of SMA wires shows improvements of healed peak fracture loads by a factor of up to 1.6, approaching the performance of the virgin material. Moreover, the repairs can be achieved with reduced amounts of healing agent. The improvements in performance were attributed mainly to the crack closure, which reduces the total crack volume and increases the crack fill factor for a given amount of healing agent. The heating of the healing agent during polymerisation increases the degree of cure of the polymerised healing agent.

4.1.3   Delaminating substrate

Because of their excellent in-plane properties and high specific strength, fibre-reinforced composites with polymeric matrices have found many uses in structural applications. Despite this success, they are particularly prone to damage from out-of-plane impact events. Although fibre damage is usually localised at the site of impact, matrix damage in the form of delaminations and transverse cracks can be more widespread. Delaminations, in particular, pose a serious issue because they can significantly reduce compressive strength [41–45] and grow in response to fatigue loading [42,46–49]. Adding to the problem, impact damage can be subsurface or barely visible, necessitating the use of expensive and time-consuming nondestructive inspection [42]. Once the damage is located, there are many repair techniques that have been proposed or are currently practiced [50–53]. Most solutions rely on resin infiltration of delaminations or composite patches to provide load transfer across the damaged region. In cases of severe damage, damaged regions are removed and replaced with a new composite material that is bonded or co-cured to the original material [50]. These repair techniques are generally time-consuming, complicated and require unhindered access. An alternative solution to manually repairing impact damage is the employment of self-healing materials. Recently, Patel *et al.* [54] have studied the autonomic self-healing of impact damage in composite materials by using a microencapsulated healing agent (DCPD liquid healing agent and paraffin wax microspheres containing 10 wt% Grubbs' catalyst), which have been successfully incorporated in a woven S2-glass-reinforced epoxy composite. Low-velocity impact tests revealed that the self-healing composite panels are able to autonomically repair impact damage. Fluorescent labelling of damage combined with image processing shows that after self-healing the total crack length per imaged cross-section is reduced by 51%.

Flexible, laminated, self-healing bladder material was also investigated to see if it could mediate the impact of small tears and punctures. Previous attempts at healing puncture damage have focused on ionomers [55] and space-filling gels [56]. A self-healing response in ionomers is initiated through the transfer of energy from a fast-moving projectile, which is typically a few millimeters in diameter. Frictional heating of the material from the passage of the projectile leads to a reorientation of the polymer chains in the ionomer. This rearrangement can, under some conditions, seal the hole generated by the projectile. However, this healing occurs only when the damaged area is heated to near the melt temperature of the material [55]. A second self-healing system, proposed by Nagaya *et al.*, utilises a water-saturated expanding gel to automatically repair tyres [56]. In this system, the polymeric gel is bonded between two layers of rubber on the inner surface of a tyre and then saturated with water. Upon puncture, the saturated gel expands and fills the puncture, sealing the leak. A 4 mm thick polymer layer is able to effectively seal nail puncture damage for typical tyre pressures of 0.25 MPa [56].

Beiermann *et al.* [57] have manufactured a three-layer, flexible self-healing material, capable of repairing puncture damage. The material used consisted of three layers: a polydimethyl siloxane composite embedded with a self-healing microcapsule system and sandwiched between two layers of PU coated nylon. A protocol was established in which samples are damaged using a hypodermic needle or a razor blade, and a successful heal is defined as the ability to reseal the damage to withstand a pressure differential across the laminate of 103 kPa (at 0.1 MPa). Healing was shown to vary significantly with microcapsule size, with the maximum healing success rate (100% successfully healed) occurring in samples with 220 µm microcapsules. Additionally, healing was found to increase with composite layer thickness and decreased with increasing puncture hole size.

Finally, fracture testing, in the form of single-edge, notched bending, shows a healing efficiency of 111% when the concentration of microcapsules and latent hardener are optimised. Some preliminary tests on epoxy-based fabric laminates containing this self-healing system demonstrated a 68% recovery of virgin interlaminar fracture toughness. Yuan *et al.* [12] reported another promising combination of healing agent and catalyst for self-healing polymer composites. The healing agents, consisting of a mixture of diglycidyl ether of bisphenol A together with a catalyst made from 1-butyl glycidyl ether, are stored in poly-UF (PUF) microcapsules that were prepared by the conventional oil-in-water emulsion process. This process of preparing the PUF microcapsules promotes long shelf-life and good chemical stability at temperatures below 238 °C. This system is still in the early developmental stages and the self-healing efficiency of the system within a composite material is yet to be tested. Kirk *et al.* [58] have used a 'physical' encapsulation of both epoxy and hardener inside nanoporous silica capsules. Once they are mixed with the polymer epoxy matrix, they are expected to withstand polymerisation because of the high aspect ratio of the channels (epoxy and hardener are deep inside the capsules). The capsules are about an order of magnitude smaller than those used before [1]. Thus, diffusion can solve the need of physical mixing of the resin and hardener, and the presence of silica capsules can also improve the mechanical properties of the composites, apart from self-healing.

4.2   Choice of the healing agent/catalyst system

The most thoroughly studied system for mechanically induced healing uses the microencapsulation of the monomers' liquid healing agent and ruthenium catalysts that will initiate the ring-opening metathesis polymerisation (ROMP) of the monomers. The polymerisation reaction is then responsible for the healing of the damaged materials. There are a number of different ways in which these components (monomers and catalysts) could be incorporated into the composite system, many of which have been explored previously.

White *et al.* [1] reported the first configuration to utilise these components, which has been widely studied. In this system, the healing agent is encapsulated in microspheres that are incorporated into the composite, and the catalyst is directly embedded into the matrix. In such a system, when a damage event occurs, the crack will propagate through the specimen, rupturing the microcapsules. The liquid healing agent will then flow through the crack *via* capillary action. Upon contact with catalyst particles in the matrix, the healing agent will undergo polymerisation to fill the crack void. In this way, crack propagation will be halted and upon curing of the healing agent, the mechanical strength of the material will be restored.

4.2.1   Healing agent

The first choice of healing agent, that is, DCPD, was made on the basis of its low cost, long shelf-life, low viscosity and volatility, and its rapid polymerisation at ambient conditions upon contact with a suitable catalyst. The chosen catalyst was the first generation Grubbs' catalyst: *bis*(tricyclohexylphosphine) benzylidene ruthenium(IV)dichloride. Grubbs' catalyst is well known for promoting olefin metathesis, showing high activity while being tolerant of a wide range of functional groups. In this groundbreaking study, White *et al.* [1] noted that the addition of microspheres and catalyst to the epoxy matrix gave an increase in the virgin fracture load of 20% as compared to the neat epoxy, illustrating the toughening effect of the filler material. The initial healing results on this type of specimen showed a maximum recovery of 75% of the virgin fracture load, with a 60% average healing efficiency. These results set the stage for a multitude of further research into this particular system. After clearly demonstrating the potential for this system to achieve fully autonomous mechanically stimulated healing, Kessler and White [59] prepared samples to test the systems for their potential use in structural composite materials. These samples introduced the microcapsules and catalyst into a CRFP to test the healing ability on the most common mode of composite failure, delamination. Following their initial results, each aspect of the two healing components has been thoroughly studied. The reactivity of the two DCPD isomers has been examined, and alternative olefin-containing healing agents have been explored. The stability, activity and costs associated with the first generation Grubbs' ruthenium catalyst have been determined and recent studies have begun investigating alternative catalysts for use in these composites.

Another similarly reactive diene-containing monomer was investigated as a possible alternative healing agent. For example, ENB reacts much faster in ROMP than DCPD, and also has a much lower freezing point than that of DCPD (15 °C). The drawback to using this monomer is that the resulting polymer is linear, and, thus has inferior mechanical properties compared to DCPD. Liu *et al.* [60] tested a system that used a blend of the two monomers as the liquid healing agent to increase the rate of polymerisation and the range of usable temperatures, while at the same time maintaining desirable mechanical properties. The polymerisation was indeed faster with the addition of ENB and could be completed at a lower catalyst loading. A sample containing a blend of DCPD/ENB showed the highest rigidity after 120 minutes of cure time, as compared to studies with the pure monomers and varied ratios of them. DCPD and ENB are presumably responsible for increases in rigidity and reactivity.

4.2.2 Ring opening metathesis polymerisation catalyst

It is commonly known that in self-healing polymers and composites, the activity of the embedded chemical catalyst within the thermosetting matrix is critical to healing efficiency. There are some issues regarding the stability and reactivity of the catalyst chosen for those systems. One of the most significant advantages of Grubbs' catalysts over other metathesis catalysts is their impressive reactivity with olefinic substrates in the presence of most common functional groups, including acids, alcohols, aldehydes, ketones and even water [61–64]. However, it has been reported that the first generation Grubbs' alkylidene can be deactivated by functional groups that coordinate strongly to the ruthenium centre. For example, the first generation Grubbs' alkylidene is known to decompose rapidly in coordinating solvents such as acetonitrile, dimethylformamide (DMF) and dimethyl sulfoxide to produce a complex mixture of ruthenium products [64].

A first generation Grubbs' catalyst can be deactivated upon prolonged exposure to air and moisture, and was also reported to diminish its reactivity upon exposure to diethylenetriamine (DETA), the agent used to cure the epoxy matrix of these composites [65]. In addition to the activity constraints of the catalyst, the particles also tended to agglomerate, which according to Kessler and White [59] led to delamination within the samples. The effective concentration of the catalyst depends on the availability of exposed catalyst, on the fracture plane, as well as the rate of dissolution of the catalyst in the healing agent. The lack of catalyst reactivity leads to partial polymerisation and poor mechanical recovery. Catalyst availability is determined by the competing rates of dissolution of the catalyst and polymerisation of the healing agent. Grubbs' catalyst exists in different crystal polymorphs and Jones *et al.* [65] reported that each kind has an effect on the dissolution kinetics and corresponding healing performance of the material. Smaller catalyst particles will have faster dissolution kinetics but will then have a larger reduction in activity due to exposure to DETA. It was reported by Taber and Frankowski [66] that Grubb's catalyst dispersed in paraffin is easily handled and retains its activity indefinitely without any special storage precautions. Rule *et al.* [67] implemented this knowledge into a system whereby the catalyst was first encapsulated in paraffin (wax encapsulation) and subsequently embedded into the epoxy matrix. This enhanced methodology proved to be quite favourable in terms of the overall effectiveness of these systems.

To utilise the wax encapsulation method for sample healing, it was first shown that the wax dissolves efficiently in the DCPD healing agent, then allowing the healing reaction to occur. The size of the resulting microspheres of the encapsulated catalyst could be controlled *via* the rate of stirring during the wax encapsulation, in the same way as in the process for encapsulation of the healing agent. The process of embedding the catalyst in wax decreased its reactivity by 9% only, which is quite low particularly when it was determined that the wax preserved up to 69% of the catalyst reactivity when it was exposed to ethylenediamine (ethylenediamine is similar in structure and reactivity to the DETA epoxy curing agent used to fabricate the composite materials [15]). Following this study, Wilson *et al.* [68] expanded upon the idea of embedding the catalyst particles in wax by studying the effect of size of these microspheres on healing efficiencies, while also introducing the catalyst into a new epoxy matrix. This new matrix material, chosen for its superior mechanical properties and performance, incorporates vinyl ester groups into the resin; this resin is then cured *via* a free radical-initiated polymerisation of the vinyl groups by an amine-peroxide species. Similar to the loss in activity resulting from the exposure of the Grubbs' catalyst to the DETA curing agent, exposure to this peroxide species would result in loss of catalytic activity.

Different ruthenium catalysts were also explored by Wilson *et al.* [69] as alternatives to using the first generation Grubbs' catalyst. Ideally, such a compound would have rapid dissolution in the healing agent, fast initiation of polymerisation, thermal stability, high processing and working temperatures and chemical stability to the matrix resins and curing agents. A first generation Grubbs' catalyst was compared with a second generation Grubbs' and second generation Hoveyda–Grubbs catalysts. The rate constant for the ROMP of DCPD in solution using each catalyst was measured. The observed rate constants were 1.45×10^{-4} s^{-1} with the first generation Grubbs' catalyst, 4.3×10^{-3} s^{-1} with the second generation Grubbs' catalyst and too fast to be measured with the Hoveyda–Grubbs catalyst. Interestingly, the first generation Grubbs' catalyst had a faster polymerisation rate in bulk when compared to the second generation Grubbs' catalyst. Furthermore, it was found that the second generation Grubbs' catalyst had the best thermal stability, and led to more efficient healing at 125 °C.

It has been reported by Jones *et al.* that the first generation catalyst maintains its activity at high temperatures up to 190 °C under nitrogen and 140 °C under air [65]. Although Grubbs' catalysts are tolerant to air and humidity and many different chemical functional groups [61], as mentioned before, the exposure of the first generation catalyst to a primary amine (e.g., DETA) has already been reported to lead to degradation of its chemical activity [3]. Rheological behaviour of ROMP-based healing agents, triggered by first or second generation Grubbs' catalyst suspended in various thermosetting resins, was investigated by Liu *et al.* [70] using an oscillatory parallel plate rheometer. From their rheological behaviour, the activity of as-received first and second generation Grubbs' catalyst was investigated when the catalyst was embedded in different thermosetting matrix resins at room and high temperatures curing agents. Both catalysts remained active and were able to initiate the ROMP reaction. It was found that the first generation catalyst was more effective than the second generation in all the epoxy matrices. The ROMP reaction rate appears to be related to the morphology and dispersion (which influence the dissolution rate) of the catalyst in the matrix resin. Gelation of healing agents initiated by first generation Grubbs' catalyst occurred faster than those triggered by the second generation catalyst. Liu *et al.* suggested that the dissolution rate of the catalyst by the healing agent is an important factor in determining the overall ROMP reaction rate *in situ*, and demonstrated that the finer, rod-like solid particles of the first generation catalyst were distributed more homogeneously throughout the cured matrix.

The stability of second generation Grubbs' alkylidenes to primary amines relative to the first generation derivatives was also investigated by Wilson *et al.* [71]. For both Grubbs' alkylidene derivatives, the tricyclohexylphosphine ligand is displaced by *n*-butyl-amine and diethylenetriamine. The result is the formation of new stable ruthenium-amine complexes, which are significantly active in ROMP and exhibited an initiation rate constant that was at least an order of magnitude greater than that of the second generation Grubbs' alkylidene from which it was synthesised.

Another aspect of this study was to investigate the inclusion of a healing agent additive, 5-norbornene-2-carboxylic acid (NCA) with the objective to promote the interaction between the matrix material and the healed polymer material. For these tests, second generation Grubbs' catalyst performed consistently well, readily polymerising the DCPD/NCA mixture to give an improved healing performance over DCPD alone. These exciting results open the door for further study of alternative healing agents and ROMP catalysts for autonomous healing systems. An alternative to the ruthenium catalysts was also recently explored, as they tend to be costly and have limited availability, and so it would not be practical for larger scale commercial applications. Tungsten (VI) catalysts were investigated by Kamphaus *et al.* [72] as a cost-effective alternative to using Grubbs' ruthenium catalyst. In these systems, the tungsten hexachloride (WC_{l6}) would act as a catalyst precursor.

4.3   Free catalyst-based epoxy/hardener and solvent encapsulation systems

Because of the stability and cost issues associated with using a catalyst in self-healing systems, alternative systems have been created that keep the same basic premise for autonomous healing.

4.3.1   Epoxy/hardener system

In recent years, systems based on self-healing of thermosetting polymer composites have attracted much attention because they represent an important class of structural materials that require long-term durability and reliability [73–82]. Dry and co-workers filled glass pipette tubes with two-part epoxy adhesives consisting of epoxy and amine hardener, and embedded them into an epoxy matrix [73,74]. To eliminate the possibility that the thick hollow glass capillaries might act as initiation for composites failure, Bleay *et al.* used hollow glass fibres possessing nearly the same diameter as the reinforcements and applied the epoxy-hardener pair as the repair agent [75]. However, filling such fine tubes with the repair species is very difficult. Jung *et al.* used polyoxymethylene-urea-walled microspheres to store an epoxide monomer to be released into cracks and rebond the cracked faces in a polyester matrix [76]. Solidification of the epoxy resin (i.e., the repair action) was triggered by the excessive amine in the composites. White *et al.* indicated that the method was not feasible as the amine groups did not retain sufficient activity [77]. Zako and Takano proposed an intelligent material system using 40% volume fraction of unmodified epoxy particles to repair microcracks and delamination damage in a glass/epoxy composite laminate [78]. By heating to 120 °C, the embedded epoxy particles (~50 μm) would melt, flow to the crack faces and repair the damage with the help of the excessive amine in the composite. Yuan *et al.* reported a self-healing epoxy composite containing epoxy-loaded PUF microcapsules [12]. The complex of $CuBr_2$ and 2-methylimidazole ($CuBr_2(2\text{-MeIm})_4$) served as latent hardener and was predissolved in the matrix during the manufacturing of the composites. Self-healing of cracks can be conducted at 130 °C as a result of the curing of the released epoxy initiated by $CuBr_2$ $(2\text{-MeIm})_4$.

It is known that self-healing based on microencapsulated healing agents offers tremendous potential for practical applications [1,17,79]. This is particularly true when mass production, long shelf-life and self-healing free of manual intervention are concerned. However, microencapsulation of hardener for epoxy healing agent is difficult, despite the fact that microcapsules containing epoxy resin are easy to synthesise by *in situ* polymerisation and interfacial copolymerisation [11,80–82]. The conventional amine-type hardeners for curing epoxy at room temperature are amphoteric and highly active, and, thus hard to be encapsulated in water or solvent by chemical methods. For example, they cannot be encapsulated by PUF under acidic conditions. Although the physical extrusion method was used to produce some hardener-loaded capsules [83,84], such as capsules containing a mixture of DETA and nonylphenol with an alginate wall and capsules containing diethylamine with a thermoplastic wall, they were not suitable for fabricating self-healing composites. The processing parameters associated with microencapsulation of epoxy resin for use as a healing agent were thoroughly investigated by Yuan *et al.* [85]. This study demonstrated the ability to control the microcapsule diameter and shell wall dimensions as well as the core content by adjusting the pH, surfactant type, time and heating rates during the microencapsulation process. Another research group has begun to investigate the use of a polythiol as a curing agent for epoxy resin, as opposed to the use of an amine-type hardener. Amine curing agents tend to be highly active, and therefore, hard to encapsulate. Yuan *et al.* [86] have reported the microencapsulation of polythiols in MF polymers, giving a curing agent that is faster as well as more stable and chemically resistant. The use of epoxy resin as a healing agent in an epoxy matrix is an important advancement to this field of microcapsule-containing self-healing composites. By replenishing voids and filling cracks with the same material as the surrounding matrix, the material as a whole remains a homogeneous, structurally uniform sample.

4.3.2   Solvent encapsulation

Earlier reports of crack healing in an epoxy resin required high-temperature conditions for healing to occur [87]. This observed healing after fracture of the virgin material was due to molecular diffusion and reaction of residual functionality during subsequent heating of the material above its glass transition temperature (T_g) [88–90]. Solvent addition is also responsible for healing, that is, ethanol and methanol were used to seal the cracks of thermoplastic polymers under high-temperature conditions [91]. Another system to be examined, which integrates an encapsulated healing agent into a polymer matrix, makes use of solvent promoted healing. It was shown by Lin *et al.* [92] that solvents could be used to assist in the healing behaviour in polymer samples, mainly during the wetting and diffusion stages. Different solvents have been used, such as methanol and/or ethanol [93,94] and carbon tetrachloride [95], and they were able to aid in the healing of polymethylmethacrylate and polycarbonate, respectively. The healing mechanism involved wetting of the polymer surface and swelling of the bulk polymer material, which led to chain interlocking across the crack plane and recovery of virgin mechanical properties. It was determined that by immersing the polymer in these solvents, the T_g of the polymer substrate could be lowered, thereby allowing for healing at or near room temperature.

Caruso *et al.* [96] transferred this research in solvent promoted healing into the realm of mechanically stimulated self-healing materials. In this system, it is the solvent that is encapsulated and embedded into the polymer matrix. Solvents were screened for their healing ability by first manually injecting them onto a crack plane of a fractured epoxy specimen. It was found that the healing efficiencies of these composites were highly correlated to solvent polarity, with the highest healing achieved by nitrobenzene, *N*-methyl pyrrolidone, dimethylacetamide, DMF and dimethyl sulfoxide. These polar aprotic solvents work well as healing agents while formamide and water, both polar protic solvents, gave no indication of healing. The encapsulation procedure of solvents proved problematic, both in using UF encapsulation and in using reverse-phase encapsulation techniques. The only solvent that was relatively easy to encapsulate was chlorobenzene, giving capsules with 160 μm average diameter. It was this encapsulated solvent that was used to demonstrate the self-healing capability of the system. Composites were fabricated with 20 wt% chlorobenzene microcapsules, which showed a maximum 82% healing efficiency. Similar composites using xylenes showed only 38% healing efficiency, and composites using hexanes showed 0% healing efficiency. This fact further demonstrated the dependence of the healing efficiency on solvent polarity (Figure 4.7).

Caruso *et al.* [99] reported two significant advances for solvent-based self-healing of epoxy materials. First, an autonomic system yielding complete recovery of fracture toughness after crack propagation was achieved by embedding microcapsules containing a mixture of epoxy monomer and solvent into an epoxy matrix. Healing with epoxy-solvent microcapsules is superior to capsules that contain solvent alone, and multiple healing events are reported for this system. Second, efficient healing is reported for new solvents, including aromatic esters, which are significantly less toxic than the previously used solvent, chlorobenzene. Preliminary ageing studies using either chlorobenzene or ethyl phenylacetate as the solvent demonstrate the stability of the epoxy-solvent system under ambient conditions

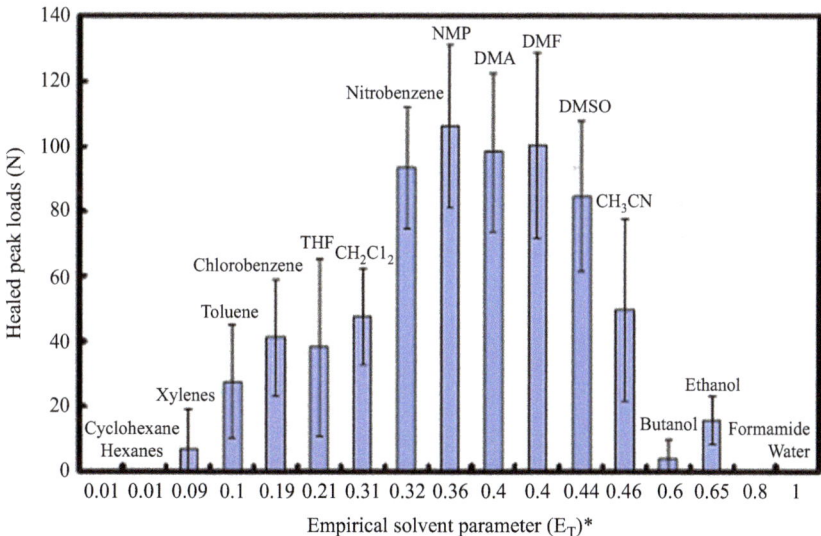

Figure 4.7   Summary of various tests for solvents exhibiting self-healing as a function of polarity. The empirical solvent parameter (E_T) is related to polarity as described in [97,98]. Error bars represent standard deviation based on 5–10 samples
[Reproduced, with permission, from [98], © 2007 American Chemical Society]

for at least 1 month. Finally, Yang *et al.* [22] have reported the fabrication of microcapsules containing reactive diisocyanate for use in self-healing polymers *via* interfacial polymerisation of PU. Isocyanates are potential catalyst-free healing agents which are capable of reacting with water in a humid or wet environment to enact healing. Microcapsules of 40–400 µm in diameter were produced by controlling the agitation rate. The microcapsules are stable with only ~10 wt% loss of isophorone diisocyanate detected after 6 months storage under ambient conditions.

4.4   Hollow glass fibres systems – two component epoxies

The development of advanced fibre-reinforced polymer(s) (FRP) to achieve performance improvements in engineering structures focuses on the exploitation of the excellent specific strength and stiffness that they offer. However, the planar nature of an FRP microstructure results in relatively poor performance under impact loading. This is an indication of their susceptibility to damage, which manifests mainly in the form of delamination (Figure 4.8). Use of reinforcing fillers of a healing material would not only add the desired strength to the system but would also allow for self-repair of any damage that occurred *via* repair materials pre-embedded into the material. Hollow glass fibres have already been shown to improve the structural performance of materials without creating areas of weakness within the composite [100,101]. These hollow fibres offer increased flexural rigidity and allow for greater custom tailoring of performance by adjusting the thickness of the walls and degree of hollowness [102,103]. By utilising hollow glass fibres in these composites, alone or in conjunction with other reinforcing fibres, it would be possible to not only

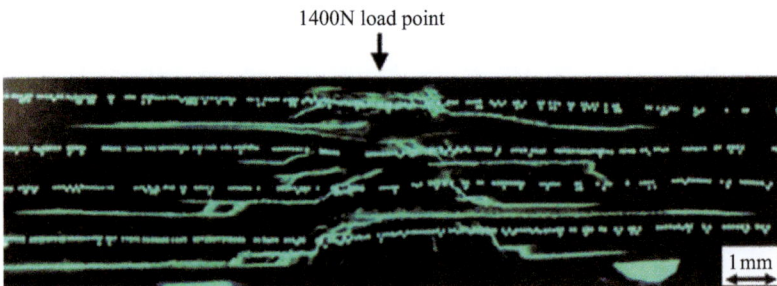

Figure 4.8   Impact damaged cross-section of GFRP laminate containing dye to visualise healing agent flow
[Reproduced, with permission, from [10], © 2007 Royal Society Publishing]

obtain the desired structural improvements, but also introduce a reservoir suitable for the containment of a healing agent [101,104]. Upon mechanical stimulus (damage inducing fracture of the fibres), this agent would 'bleed' into the damage site to initiate repair, not unlike biological self-healing mechanisms [105,106].

The first systems, which were investigated by Dry [107] and Williams *et al.* [108], proved that the proposed architecture for releasing chemicals from repair fibres was possible. Afterwards, the authors used cyanoacrylate, ethyl cyanoacrylate [109,110], and methyl methacrylate [111,112] as healing agents to heal cracks in concrete. This methodology was then transferred to polymer composite materials by Motuku *et al.* [113]. The healing agents contained within the glass fibres have either been a one-part adhesive, such as cyanoacrylate, or a two-part epoxy system containing both a resin and a hardener, where both are loaded in perpendicular fibres or one is embedded into the matrix and the other inside the fibres [92].

One of the initial challenges encountered when creating this type of self-healing system is the development of a practical technique for filling the hollow glass fibres with repair agent. When approaching this problem, the dimensions of the glass fibre itself must be considered, including diameter, wall thickness, hollowness of the fibre as well as the viscosity and healing kinetics of the repair agent. Bleay *et al.* [75] were among the first to develop and implement a fibre filling method involving 'capillary action' that is assisted by vacuum, which is now the commonly used process. The chosen glass fibre should be also evaluated for its capacity to survive the composite manufacturing process without breakage, while still possessing its ability to rupture during a damage event in order to release the required healing agent. Motuku *et al.* [113] have clearly determined that hollow glass fibres were best suited for this kind of application, as opposed to polymer tubes or those made of metal, which often did not provide controlled fracture upon impact damage. Hucker *et al.* [103] have shown that hollow glass fibres of a larger diameter, from 30 to 60 µm, offered an increased compressive strength while giving a larger volume of repair agent to be stored within the composite. The second important parameter to investigate was the capacity of the healing agent to adequately reach the site of damage and subsequently undergo healing. This mechanism will obviously depend upon the viscosity of the healing material, as well as on the kinetics of the repair process. For example, the cyanoacrylate system studied by Bleay *et al.* [75] was shown to restore mechanical strength to damaged specimens, but also caused significant problems by curing upon contact with the opening of the fibre, which prevented the healing agent from reaching the site of damage in the sample. Various studies [92,101,105,106,113] have used liquid dyes inside the composites to act as a damage detection mechanism providing a visible indication of the damage site, while allowing a clear evaluation of the flow of the healing agents to those sites. Finally, the third parameter to optimise, is the concentration of healing fibres within the matrix, their spatial distribution and the dimensions of the specimen, which all have a direct effect on the mechanical properties of the resulting composite material. As demonstrated by Jang *et al.* [114], the stacking sequence of the fibres within the composite plays a role in inhibiting plastic deformation and delamination, and will also affect the response to an impact damage event. In order to maintain high mechanical properties, repaired fibres need to be adequately spaced within the composite. Motuku *et al.* [113] have shown that thicker composites perform better in the healing process. These parameters, however, will depend upon the choice of the dimension of the fibre and on the choice of chemical used for the healing agent, and so optimisation will depend on the specifications of the system being studied.

Until recently, the majority of the research done on self-healing hollow fibre composites have focused on demonstrating the feasibility of such a concept for self-repair, and have reported qualitatively on the healing capacity of the studied systems. Recently, numerous works have reported quantitatively on the mechanical properties associated with the healing of the materials. The inclusion of hollow glass fibres into a composite system was shown by Williams *et al.* [104] and Trask and Bond [107] to give an initial reduction in the strength of the material, either by 16% in glass fibre reinforced polymer (GFRP) composites or by 8% in CFRP composites. These 'self-repairing' composites were shown to recover 100% of the virgin strength for GFRP and 97% of the virgin strength for CFRP. In both cases the composite materials were subjected to a heat treatment to aid in the delivery of the resin to the damaged area as well as in the curing of the healing agent. Recently, Williams *et al.* [108] have considered the development of autonomic self-healing within a CFRP. They have demonstrated the significant strength recovery (>90%) when a resin filled, hollow glass fibre system was distributed at specific interfaces within a laminate, minimising the reduction in mechanical properties whilst maximising the efficiency of the healing event.

4.5   Microvascular networks systems

Once the biomimetic approach to the incorporation of microvascular networks into a composite material is understood, a new class of self-healing materials can be developed. As outlined by Stroock and Cabodi [115], these microvascular networks can be created *via* soft lithographic methods, in which all microchannels can be fabricated at the same time, or through direct write methods, which are more suited for building three-dimensional (3D) microchannel structures. The microchannels can then be filled with a liquid healing agent. Upon damage to the channel, the fluid will be released into the composite and subsequently undergo healing. As reported by Murphy and Wudl [15], complex microvascular networks are widely observed in biological systems, such as leaf venation [116–119] and blood vascularisation [120–122]. Indeed, in the latter case, the human circulatory system is comprised of vessels of varying diameter and length: arteries, veins and capillaries. These vessels function together in a branched system to supply blood to all points in the body simultaneously. However, due to their complex architecture, replication of these microvascular systems remains a significant challenge for those pursuing synthetic analogues. Various techniques, including soft lithography [123–125], laser ablation [126,127] and direct-write assembly [128] have been already used to create planar and 3D microvascular networks. Recent research has been focused on the fabrication of a microvascular network containing self-healing materials, with the purpose of developing a biomimetic material. One of the main advantages, of using both the glass fibre and microcapsule systems, is their ability to heal the same location in the material more than once. This is a very attractive option, because quite often, a second fracture event will occur along the plane of the initial crack. By providing a material with a quasi-continuous flow of healing agent, numerous healing cycles can be achieved.

Toohey *et al.* [129] described one of the first of these types of composite material. Authors reported self-healing systems which are capable of autonomously repairing repeated damage events. They reported the system, in which bio-inspired coating-substrate delivered healing agent to crack in a polymer coating *via* a 3D microvascular network [128]. Crack damage in the epoxy coating was healed repeatedly. As mentioned above, this approach opens new possibilities for continuous delivery of healing agents for self-repair as well as other active species for additional functionality. This system utilised the healing combination of liquid DCPD as the healing agent, and solid Grubbs' catalyst to initiate ROMP polymerisation of the DCPD. In the reported work, catalyst was incorporated into a 700 μm thick epoxy coating that was applied to the top surface of the microvascular substrate, and the 200 μm wide channels were successfully filled with DCPD and then sealed. The system achieved a peak healing efficiency up to 70% with 10 wt% catalyst in the top coating, and was able to demonstrate healing for up to seven cycles. The amount of catalyst in the top epoxy layer did not affect the average healing efficiency per cycle but rather limited how many cycles of testing and healing could be performed successfully. Indeed, once all of the catalyst has been used, the healing ceased due to depletion of catalyst in the crack plane, even with a continuous supply of monomer.

To overcome this limitation, Toohey *et al.* in 2007 modified this design by photo-lithographically patterning four isolated regions within the embedded microvascular network [129]. They reported on the repeated healing of crack damage in a polymeric coating through delivery of two-part epoxy, and healing chemistry *via* multiple microvascular networks embedded in isolation within a polymeric substrate. The authors first created a continuous, interconnected microvascular network using the direct-write assembly method. They then isolated multiple networks by infilling the network with a photo-curable resin and selectively photo-polymerising thin parallel sections of these resin-filled microchannels. Epoxy resin and amine-based curing agents are transported to the crack plane through two sets of independent vascular networks embedded within a ductile polymer substrate beneath the coating. The two reactive components remain isolated and stable in the vascular networks, until crack formation occurs in the coating under a mechanical load. Both healing components are wicked by capillary forces into the crack plane, where they react and effectively bond the crack. Several epoxy and curing agent combinations were evaluated for their suitability in microvascular-based autonomic systems. Healing efficiencies of over 60% for up to 16 intermittent healing events out of 23 cycles were successfully achieved.

In a related work, Williams *et al.* published their version of a microvascular configuration containing mechanically stimulated healable material in the form of sandwich structure composite configurations that contain either single [130] or dual [131] fluidic networks. In the single network design, sandwich structures use high performing skin materials, such as glass or carbon fibre composites, separated by a lightweight core to obtain a material with a very high specific flexural stiffness. A vascular network incorporated into a sandwich structure would address the larger damage volume expected of these systems, as well as allowing for multiple healing events to occur. Samples were fabricated with channels containing a healing agent, which had a negligible effect on the mechanical properties of the composite. Rupture of the vessels released the healing fluid, filling the void that formed as a result of impact damage on the sample. Initial tests were run on samples containing premixed resin and hardener to demonstrate the healing capability of the system. Indeed, these samples showed consistent and complete recovery of compressive stress at failure after impact damage. In their dual network design, significant recovery is observed when samples are infiltrated with pressurized, unmixed dual fluids [131] (Figure 4.9).

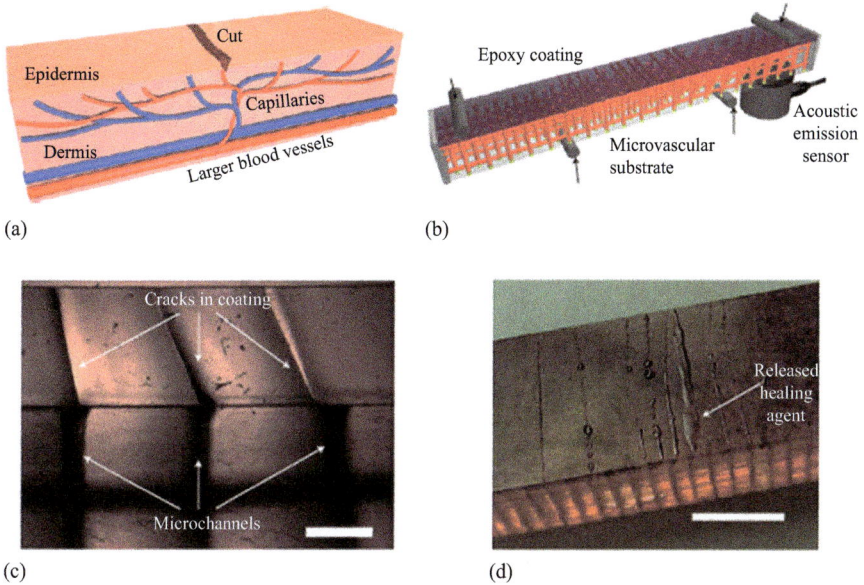

Figure 4.9 *Self-healing materials with 3D microvascular networks. (a) Schematic diagram of a capillary network in the dermis layer of skin with a cut in the epidermis layer; (b) schematic diagram of the self-healing structure composed of a microvascular substrate and a brittle epoxy coating containing embedded catalyst in a four-point bending configuration monitored with an acoustic-emission sensor; (c) high-magnification cross-sectional image of the coating showing that cracks, which initiate at the surface, propagate towards the microchannel openings at the interface (scale bar = 0.5 mm) and (d) optical image of self-healing structure after cracks are formed in the coating (with 2.5 wt% catalyst), revealing the presence of excess healing fluid on the coating surface (scale bar = 5 mm)* [Reproduced, with permission, from [129], © 2007 Nature Publishing Group]

4.6 Self-healing coatings for metallic structures

The large economic impact of corrosion of metallic structures is a very important issue. Generally, rapid field-specific testing is done when material failure is observed. Despite intense research and development in corrosion protection coatings of metals and alloys, the actual performance results are not always satisfactory. Furthermore, the development of coatings to protect and prolong the service life of the infrastructure is still a big challenge, because of the wide variations in environmental conditions. Therefore, in order to improve the equipment service prediction capabilities of infrastructure, it is indispensable to develop new state-of-the-art smart/self-healing coating formulations for corrosion inhibition. In this context, autonomic healing materials respond without external intervention to environmental stimuli, and have great potential for advanced engineering systems [1,5,13,24,132–144]. Self-healing coatings, which autonomically repair and prevent corrosion of the underlying substrate, are of particular interest. Notably, the worldwide cost of corrosion has been estimated to be nearly US\$ 300 billion per year [145]. Recent studies on self-healing polymers have demonstrated repair of bulk mechanical damage as well as dramatic increases in the fatigue life. The majority of these systems, however, have serious chemical and mechanical limitations, preventing their use as coatings. Polymer coating systems are traditionally applied to a metal surface to provide a dense barrier against the corrosive species. Cathodic protection is used for many applications in addition to coatings to protect the metal structures from corrosive attack when the coating is damaged. Thus, self-healing coatings are considered as an alternative route for efficient anti-corrosion protection while maintaining a low demand in cathodic protection.

Cho *et al.* [146] have explored two self-healing coating approaches, starting from the siloxane-based materials system. In the first approach, the catalyst was microencapsulated, and the siloxanes are used as phase-separated droplets. In the second approach, the siloxanes were also encapsulated and dispersed in the coating matrix. Encapsulation of both phases (the catalyst and the healing agent) is advantageous in cases where the matrix can react with the healing agent. Aramaki [147,148] has prepared a highly protective and self-healing film of organosiloxane polymer containing sodium silicate and cerium nitrate ($Ce(NO_3)_3$), on a zinc electrode previously treated in a $Ce(NO_3)_3$ solution. The film was examined by polarisation measurement of the electrode in an aerated 0.5 M NaCl solution after the electrode was scratched and immersed in the solution for several hours. The self-healing mechanism of the film was investigated after being scratched and immersed in the NaCl solution. A passive film was formed on the scratched surface, resulting in suppression of pitting corrosion at the scratch. More recently, Aramaki and Shimura [149,150] prepared an ultra-thin two-dimensional polymer coating, on a passivated iron electrode, which was subsequently healed in sodium nitrate. The localised corrosion was markedly prevented by coverage with the polymer coating and the healing treatment in 0.1 M $NaNO_3$. Indeed, evidence of protection of iron from corrosion in 0.1 M NaCl was observed.

The development of effective corrosion-inhibitor coatings for the prevention of corrosion-initiation and suppression of galvanic activity of metals and alloys has always been a challenging problem. Recently the concerted efforts of researchers at the US Army Engineering Research and Development Center at the Construction Engineering Research Laboratory and at other facilities [151–153] have led to the development of self-healing corrosion inhibitors to reduce or prevent corrosion of metal hardware. Previously, heavy metal-based epoxy primer pretreatment systems [154,155], including quaternary ammonium salt-based and multifunctional microencapsulated corrosion-inhibitor systems [153], have demonstrated corrosion protection performance of metals and alloys. The studies have demonstrated that the damaged film area on otherwise corroded panels experienced little lifting and blisters of the film because of the presence of microcapsules at the scribes. Mehta and Bogere [156] have evaluated the smart/self-healing microencapsulated inhibitor incorporated in epoxy primer before painting on a steel surface for its corrosion protection effectiveness on exposure to ASTM D5894 [157] electrolyte in a laboratory and a natural tropical sea-shore environment. Their results have clearly indicated that, first, the active components were successfully released, and second, the undamaged surface film has demonstrated excellent corrosion-inhibition performance as reflected by both visual inspection and electrochemical impedance spectroscopy experimental data.

In summary, the results presented in this section should provide an understanding of the fundamental material–property relationships of smart inhibitor coatings. And, thus, should facilitate the development of optimised paint compositions in order to extend the useful service life of steel-infrastructure applications.

References

[1]   S.R. White, N.R. Sottos, P.H. Geubelle, *et al.*, *Nature*, 2001, **409**, 6822, 794.

[2]   J.M. Asua, *Progress in Polymer Science*, 2002, **27**, 7, 1283.

[3]   E.N. Brown, N.R. Sottos and S.R. White, *Experimental Mechanics*, 2002, **42**, 4, 372.

[4]   M.R. Kessler, N.R. Sottos and S.R. White, *Composites Part A: Applied Science and Manufacturing*, 2003, **34**, 8, 743.

[5]   E.N. Brown, S.R. White and N.R. Sottos, *Journal of Materials Science*, 2004, **39**, 5, 1703.

[6]   E.N. Brown, S.R. White and N.R. Sottos, *Composites Science and Technology*, 2005, **65**, 15–16, 2466.

[7] E.N. Brown, S.R. White and N.R. Sottos, *Composites Science and Technology*, 2005, **65**, 15–16, 2474.

[8] A.S. Jones, J.D. Rule, J.S. Moore, N.R. Sottos and S.R. White, *Journal of the Royal Society –Interface*, 2007, **4**, 13, 395.

[9] E.N. Brown, M.R. Kessler, N.R. Sottos and S.R. White, *Journal of Microencapsulation*, 2003, **20**, 6, 719.

[10] S. Cosco, V. Ambrogi, P. Musto and C. Carfagna, *Macromolecular Symposia*, 2006, **234**, 1, 184.

[11] L. Yuan, G. Liang, J.Q. Xie, L. Li and J. Guo, *Polymer*, 2006, **47**, 15, 5338.

[12] L. Yuan, G.Z. Liang, J.Q. Xie, L. Li and J. Guo, *Journal of Materials Science*, 2007, **42**, 12, 4390.

[13] T. Yin, M.Z. Rong, M.Q. Zhang and G.C. Yang, *Composites Science and Technology*, 2007, **67**, 2, 201.

[14] B.J. Blaiszik, M.M. Caruso, D.A. McIlroy, J.S. Moore, S.R. White and N.R. Sottos, *Polymer*, 2009, **50**, 4, 990.

[15] E.B. Murphy and F. Wudl, *Progress in Polymer Science*, 2010, **35**, 1–2, 223.

[16] M.W. Keller and N.R. Sottos, *Experimental Mechanics*, 2006, **46**, 6, 725.

[17] J.D. Rule, N.R. Sottos and S.R. White, *Polymer*, 2007, **48**, 12, 3520.

[18] B.J. Blaiszik, N.R. Sottos and S.R. White, *Composites Science and Technology*, 2008, **68**, 3–4, 978.

[19] S-J. Park, Y-S. Shin and J-R. Lee, *Journal of Colloid and Interface Science*, 2001, **241**, 2, 502.

[20] L. Yuan, G-Z. Liang, J-Q. Xie and S-B. He, *Colloid and Polymer Science*, 2007, **285**, 7, 781.

[21] Y.C. Yuan, M.Z. Rong, M.Q. Zhang, J. Chen, G.C. Yang and X.M. Li, *Macromolecules*, 2008, **41**, 14, 5197.

[22] J. Yang, M.W. Keller, J.S. Moore, S.R. White and N.R. Sottos, *Macromolecules*, 2008, **41**, 24, 9650.

[23] X. Liu, X. Sheng, J.K. Lee and M.R. Kessler, *Macromolecular Materials and Engineering*, 2009, **294**, 6–7, 389.

[24] J.Y. Lee, G.A. Buxton and A.C. Balazs, *Journal of Chemical Physics*, 2004, **121**, 11, 5531.

[25] Z. Zhang, R. Saunders and C.R. Thomas, *Journal of Micro-encapsulation*, 1999, **16**, 1, 117.

[26] G. Sun and Z. Zhang, *Journal of Microencapsulation*, 2001, **18**, 5, 593.

[27] G. Sun and Z. Zhang, *International Journal of Pharmaceutics*, 2002, **242**, 1–2, 307.

[28] L-Y. Chu, R. Xie, J-H. Zhu, W-M. Chen, T. Yamaguchi and S-I. Nakao, *Journal of Colloid and Interface Science*, 2003, **265**, 1, 187.

[29] S. Zywicki and A. Bartkowiak, *Przeglad Papierniczy*, 2005, **61**, 5, 261.

[30] M.A. White, *Journal of Chemical Education*, 1998, **75**, 9, 1119.

[31] S.K. Ghosh, in *Functional Coatings by Polymer Encapsulation*, Ed., S.K. Ghosh, Wiley-VCH Verlag GmbH & Co. KgaA, Weinheim, Germany, 2006, p. 15.

[32] M. Ozdemir and T. Cevik, in *Encapsulation and Controlled Release Technologies in Food Systems*, Ed., J.M. Lakkis, Blackwell Publishers, Oxford, 2007, p. 201.

[33] C. Andersson, L. Järnström, A. Fogden, *et al.*, *Packaging Technology and Science*, 2009, **22**, 5, 275.

[34] S.D. Mookhoek, B.J. Blaiszik, H.R. Fischer, N.R. Sottos, S.R. White and S. Zwaaga, *Journal of Materials Chemistry*, 2008, **18**, 44, 5390.

[35] E. Haddad, R.V. Kruzelecky, W. P. Liu, and S. V. Hoa, *Innovative Self-repairing of Space CFRP Structures and Kapton Membranes – A Step Towards Completely Autonomous Health Monitoring and Self-healing*, Final Report, Contract No: CSA # 28-7005715, Canadian Space Agency, Quebec, Canada, 2009.

[36] M.W. Keller, S.R. White and N.R. Sottos, *Polymer*, 2008, **49**, 13–14, 3136.

[37] J. Schrooten, V. Michaud, J. Parthenios, *et al.*, *Materials Transactions*, 2002, **43**, 5, 961.

[38] K.A. Tsoi, J. Schrooten and R. Stalmans, *Materials Science and Engineering A*, 2004, **368**, 1–2, 286.

[39] C.A. Rogers, C. Liang and S. Li, *Proceedings of the AIAA/ASME/ASCE/AMS/ASC 32nd Conference – Structures, Structural Dynamics and Materials Conference*, Baltimore, MD, 1991, p. 1190.

[40] E.L. Kirkby, J.D. Rule, V.J. Michaud, N.R. Sottos, S.R. White and J-A.E. Månson, *Advanced Functional Materials*, 2008, **18**, 15, 2253.

[41] G. Zhou and L.J. Greaves, *Impact Behavior of Fibre-reinforced Composite Materials*, Eds., S.R. Reid and G. Zhou, Woodhead Publishing Ltd, Cambridge and CRC Press LLC, Boca Raton, FL, 2000, p. 133.

[42] A.A. Baker, R. Jones and R.J. Callinan, *Composite Structures*, 1985, **4**, 1, 15.

[43] J.C. Prichard and P.J. Hogg, *Composites*, 1990, **21**, 6, 503.

[44] F.J. Guild, P.J. Hogg and J.C. Prichard, *Composites*, 1993, **24**, 4, 333.

[45] Y. Xiong, C. Poon, P.V. Straznicky and H. Vietinghoff, *Composite Structures*, 1995, **30**, 4, 357.

[46] A.S. Chen, D.P. Almond and B. Harris, *International Journal of Fatigue*, 2002, **24**, 2-4, 257.

[47] D.Y. Konishi and W.R. Johnston, *Proceedings of the ASTM 5th Composite Materials Conference: Testing and Design*, New Orleans, LA, 1978, p. 597.

[48] M. Mitrovic, H.T. Hahn, G.P. Carman and P. Shyprykevich, *Composites Science and Technology*, 1999, **59**, 14, 2059.

[49] R.L. Ramkumar, *Proceedings of the ASTM Long-term Behavior of Composites Conference*, Williamsburg, VA, 1983, p. 116.

[50] S.H. Myhre and J.D. Labor, *Journal of Aircraft*, 1981, **18**, 7, 546.

[51] R.B. Heslehurst, *SAMPE Journal*, 1997, **33**, 5, 11.

[52] R.B. Heslehurst, *SAMPE Journal*, 1997, **33**, 6, 16.

[53] L. Dorworth, G. Gardiner and A. Training, *Journal of Advanced Materials*, 2007, **39**, 4, 3.

[54] A.J. Patel, N.R. Sottos, E.D. Wetzel and S.R. White, *Composites Part A: Applied Science and Manufacturing*, 2010, **41**, 3, 360.

[55] S.J. Kalista Jr., T.C. Ward and Z. Oyetunji, *Mechanics of Advanced Materials and Structure*, 2007, **14**, 5, 391.

[56] K. Nagaya, S. Ikai, M. Chiba and X. Chao, *JMSE International Journal Series C*, 2006, **49**, 2, 379.

[57] B.A. Beiermann, M.W. Keller and N.R. Sottos, *Smart Materials and Structures*, 2009, **18**, 8, 085001.

[58] J.G. Kirk, S. Naik, J.C. Moosbrugger, D.J. Morrison, D. Volkov and I. Sokolov, *International Journal of Fracture*, 2009, **159**, 101.

[59] M.R. Kessler and S.R. White, *Composites Part A: Applied Science and Manufacturing*, 2001, **32**, 5, 683.

[60] X. Liu, J.K. Lee, S.H. Yoon and M.R. Kessler, *Journal of Applied Polymer Science*, 2006, **101**, 3, 1266.

[61] T.M. Trnka and R.H. Grubbs, *Accounts of Chemical Research*, 2001, **34**, 1, 18.

[62] A. Fürstner, *Angewandte Chemie International Edition*, 2000, **39**, 17, 3012.

[63] R.H. Grubbs and S. Chang, *Tetrahedron*, 1998, **54**, 18, 4413.

[64] R.H. Grubbs, *Handbook of Metathesis*, Wiley-VCH Verlag GmbH & Co. KgaA, Weinheim, Germany, 2003.

[65] A.S. Jones, J.D. Rule, J.S. Moore, S.R. White and N.R. Sottos, *Chemistry of Materials*, 2006, **18**, 5, 1312.

[66] D.F. Taber and K.J. Frankowski, *The Journal of Organic Chemistry*, 2003, **68**, 15, 6047.

[67] J.D. Rule, E.N. Brown, N.R. Sottos, S.R. White and J.S. Moore, *Advanced Materials*, 2005, **17**, 2, 205.

[68] G.O. Wilson, J.S. Moore, S.R. White, N.R. Sottos and H.M. Andersson, *Advanced Functional Materials*, 2008, **18**, 1, 44.

[69] G.O. Wilson, M.M. Caruso, N.T. Reimer, S.R. White, N.R. Sottos and J.S. Moore, *Chemistry of Materials*, 2008, **20**, 10, 3288.

[70] X. Liu, X. Sheng, J.K. Lee, M.R. Kessler and J.S. Kim, *Composites Science and Technology*, 2009, **69**, 13, 2102.

[71] G.O. Wilson, K.A. Porter, H. Weissman, S.R. White, N.R. Sottos and J.S. Moore, *Advanced Synthesis and Catalysis*, 2009, **351**, 11–12, 1817.

[72] J.M. Kamphaus, J.D. Rule, J.S. Moore, N.R. Sottos and S.R. White, *Journal of the Royal Society – Interface*, 2008, **5**, 18, 95.

[73] C. Dry and N.R. Sottos, *SPIE Proceedings Volume 1916, Smart Structures and Materials: Smart Materials*, Ed., V.K. Varadan, Bellingham, WA, 1993, p. 438.

[74] C. Dry, *Composite Structures*, 1996, **35**, 3, 263.

[75] S.M. Bleay, C.B. Loader, V.J. Hawyes, L. Humberstone and P.T. Curtis, *Composites Part A: Applied Science and Manufacturing*, 2001, **32**, 12, 1767.

[76] D. Jung, A. Hegeman, N.R. Sottos, P.H. Geubelle and S.R. White, *The American Society for Mechanical Engineers, Materials Division*, 1997, **80**, 265.

[77] S.R. White, N.R. Sottos, P.H. Geubelle, *et al.*, inventors; University of Illinois, assignee; US 6858659, 2005.

[78] M. Zako and N. Takano, *Journal of Intelligent Material Systems and Structures*, 1999, **10**, 10, 836.

[79] N.R. Sottos, S.R. White and I. Bond, *Journal of the Royal Society – Interface*, 2007, **4**, 13, 347.

[80] S. Matsuo, I. Usami, M. Kurihara and K. Nakashima, inventors; Three Bond Co, assignee; EP 0543675A1, 1993.

[81] R.L. Hart, D.E. Work and C.E. Davis, inventors; Capsulated Systems, Inc., assignee; US 4536524, 1985.

[82] D.S. Xiao, M.Z. Rong and M.Q. Zhang, *Polymer*, 2007, **48**, 16, 4765.

[83] C.E. Schuetze, inventor; Amp Inc., assignee; US 3396117, 1968.

[84] C.R. Goldsmith, inventor; Phillips Petroleum Co., assignee; US 3791980, 1974.

[85] L. Yuan, A. Gu and G. Liang, *Materials Chemistry and Physics*, 2008, **110**, 2–3, 417.

[86] Y.C. Yuan, M.Z. Rong and M.Q. Zhang, *Polymer*, 2008, **49**, 10, 2531.

[87] J.O. Outwater and D.J. Gerry, *Journal of Adhesion*, 1969, **1**, 4, 290.

[88] K. Jud and H.H. Kausch, *Polymer Bulletin*, 1979, **1**, 10, 697.

[89] R.P. Wool and K.M. O'Connor, *Journal of Applied Physics*, 1981, **52**, 10, 5953.

[90]   J. Raghavan and R.P. Wool, *Journal of Applied Polymer Science*, 1999, **71**, 5, 775.

[91]   J-S. Shen, J.P. Harmon and S. Lee, *Journal of Materials Research*, 2002, **17**, 6, 1335.

[92]   C.B. Lin, S. Lee and K.S. Liu, *Polymer Engineering and Science*, 1990, **30**, 21, 1399.

[93]   P-P. Wang, S. Lee and J.P. Harmon, *Journal of Polymer Science, Part B: Polymer Physics Edition*, 1994, **32**, 7, 1217.

[94]   H-C. Hsieh, T-J. Yang and S. Lee, *Polymer*, 2001, **42**, 3, 1227.

[95]   T. Wu and S. Lee, *Journal of Polymer Science, Part B: Polymer Physics Edition*, 1994, **32**, 12, 2055.

[96]   M.M. Caruso, D.A. Delafuente, V. Ho, N.R. Sottos, J.S. Moore and S.R. White, *Macromolecules*, 2007, **40**, 25, 8830.

[97]   C. Reichardt, *Solvents and Solvent Effects in Organic Chemistry*, Wiley-VCH, New York, NY, 1988, p. 407.

[98]   M. M. Caruso, D. A. Delafuente, V. Ho, N. R. Sottos, J.S. Moore, and S.R. White, *Macromolecules*, 2007, **40**, 8830.

[99]   M.M. Caruso, B.J. Blaiszik, S.R. White, N.R. Sottos, and J.S. Moore, *Advanced Functional Materials*, 2008, **18**, 13, 1898.

[100] M.J. Hucker, I.P. Bond, S. Haq, S. Bleay and A. Foreman, *Journal of Materials Science*, 2002, **37**, 2, 309.

[101] R.S. Trask, G.J. Williams and I.P. Bond, *Journal of the Royal Society – Interface*, 2007, **4**, 13, 363.

[102] M. Hucker, I. Bond, A. Foreman and J. Hudd, *Advanced Composites Letters*, 1999, **8**, 4, 181.

[103] M. Hucker, I. Bond, S. Bleay and S. Haq, *Composites Part A: Applied Science and Manufacturing*, 2003, **34**, 10, 927.

[104] G. Williams, R. Trask and I. Bond, *Composites Part A: Applied Science and Manufacturing*, 2007, **38**, 6, 1525.

[105] J.W.C. Pang and I.P. Bond, *Composites Science and Technology*, 2005, **65**, 11–12, 1791.

[106] J.W.C. Pang and I.P. Bond, *Composites Part A: Applied Science and Manufacturing*, 2005, **36**, 2, 183.

[107] R.S. Trask and I.P. Bond, *Smart Materials and Structures*, 2006, **15**, 3, 704.

[108] G.J. Williams, I.P. Bond and R.S. Trask, *Composites Part A: Applied Science and Manufacturing*, 2009, **40**, 9, 1399.

[109] V.C. Li, Y.M. Lim and Y-W. Chan, *Composites Part B: Engineering*, 1998, **29**, 6, 819.

[110] C. Dry, *International Journal of Modern Physics B*, 1992, **6**, 15–16, 2763.

[111] C. Dry and W. McMillan, *Smart Materials and Structures*, 1996, **5**, 3, 297.

[112] C. Dry, *Smart Materials and Structures*, 1994, **3**, 2, 118.

[113] M. Motuku, U.K. Vaidya and G.M. Janowski, *Smart Materials and Structures*, 1999, **8**, 5, 623.

[114] B.Z. Jang, L.C. Chen, L.R. Hwang, J.E. Hawkes and R.H. Zee, *Polymer Composites*, 1990, **11**, 3, 144.

[115] A.D. Stroock and M. Cabodi, *MRS Bulletin*, 2006, **31**, 2, 114.

[116] G.B. West, J.H. Brown and B.J. Enquist, *Nature*, 1999, **400**, 6745, 664.

[117] A. Roth-Nebelsick, D. Uhl, V. Mosbrugger and H. Kerp, *Annals of Botany*, 2001, **87**, 5, 553.

[118] N.M. Holbrook and M.A. Zwieniecki, Eds., *Vascular Transport in Plants*, Elsevier Academic Press, Burlington, MA, 2005.

[119] L. Sack and K. Frole, *Ecology*, 2006, **87**, 2, 483.

[120] G.B. West, J.H. Brown and B.J. Enquist, *Science*, 1997, **276**, 5309, 122.

[121] B. Sapoval, M. Filoche and E.R. Weibel, *Proceedings of the National Academy of Sciences of the United States of America*, 2002, **99**, 16, 10411.

[122] R.K. Jain, *Science*, 2005, **307**, 5706, 58.

[123] N.W. Choi, M. Cabodi, B. Held, J.P. Gleghorn, L.J. Bonassar and A.D. Stroock, *Nature Materials*, 2007, **6**, 11, 908.

[124] M.K. Runyon, B.L. Johnson-Kerner, C.J. Kastrup, T.G. Van Ha and R.F. Ismagilov, *Journal of the American Chemical Society*, 2007, **129**, 22, 7014.

[125] J.M. Higgins, D.T. Eddington, S.N. Bhatia and L. Mahadevan, *Proceedings of the National Academy of Sciences of the United States of America*, 2007, **104**, 51, 20496.

[126] D. Lim, Y. Kamotani, B. Cho, J. Mazumder and S. Takayama, *Lab on a Chip*, 2003, **3**, 4, 318.

[127] D.H. Kam and J. Mazumder, *Journal of Laser Applications*, 2008, **20**, 3, 185.

[128] D. Therriault, S.R. White and J.A. Lewis, *Nature Materials*, 2003, **2**, 4, 265.

[129] K.S. Toohey, N.R. Sottos, J.A. Lewis, J.S. Moore and S.R. White, *Nature Materials*, 2007, **6**, 8, 581.

[130] H.R. Williams, R.S. Trask and I.P. Bond, *Smart Materials and Structures*, 2007, **16**, 4, 1198.

[131] H.R. Williams, R.S. Trask and I.P. Bond, *Composites Science and Technology*, 2008, **68**, 15-16, 3171.

[132] M.W. Keller, S.R. White and N.R. Sottos, *Advanced Functional Materials*, 2007, **17**, 14, 2399.

[133] D.G. Shchukin and H. Möhwald, *Small*, 2007, **3**, 6, 926.

[134] X.X. Chen, M.A. Dam, K. Ono, *et al.*, *Science*, 2002, **295**, 5560, 1698.

[135] X.X. Chen, F. Wudl, A.K. Mal, H.B. Shen and S.R. Nutt, *Macromolecules*, 2003, **36**, 6, 1802.

[136] F.R. Kersey, D.M. Loveless and S.L. Craig, *Journal of the Royal Society – Interface*, 2007, **4**, 13, 373.

[137] P. Cordier, F. Tournilhac, C. Soulié-Ziakovic and L. Leibler, *Nature*, 2008, **451**, 7181, 977.

[138] S. Gupta, Q.L. Zhang, T. Emrick, A.C. Balazs and T.P. Russell, *Nature Materials*, 2006, **5**, 3, 229.

[139] R. Verberg, A.T. Dale, P. Kumar, A. Alexeev and A.C. Balazs, *Journal of the Royal Society – Interface*, 2007, **4**, 13, 349.

[140] C.S. Coughlin, A.A. Martinelli and R.F. Boswell, *Polymeric Materials Science and Engineering*, 2004, **91**, 472.

[141] S.J. Kalista, Jr., and T.C. Ward, *Journal of the Royal Society – Interface*, 2007, **4**, 13, 405.

[142] D.V. Andreeva, D. Fix, H.M. Mohwald and D.G. Shchukin, *Advanced Materials*, 2008, **20**, 14, 2789.

[143] J.W.C. Pang and I.P. Bond, *Composites Part A: Applied Science and Manufacturing*, 2005, **36**, 2, 183.

[144] S.H. Cho, H.M. Andersson, S.R. White, N.R. Sottos and P.V. Braun, *Advanced Materials*, 2006, **18**, 8, 997.

[145] G.H. Koch, M.P.H. Brongers, N.G. Thompson, Y.P. Virmani and J.H. Payer, *Corrosion Costs and Preventive Strategies in the United States*, Publication No. FHWA-RD-01-156, US Department of Transportation, Federal Highway Administration, Washington, DC, 2001.

[146] S.H. Cho, S.R. White and P.V. Braun, *Advanced Materials*, 2009, **21**, 6, 645.

[147] K. Aramaki, *Corrosion Science*, 2002, **44**, 7, 1621.

[148] K. Aramaki, *Corrosion Science*, 2003, **45**, 1, 199.

[149] K. Aramaki and T. Shimura, *Corrosion Science*, 2010, **52**, 4, 1464.

[150] K. Aramaki and T. Shimura, *Corrosion Science*, 2010, **52**, 1, 1.

[151] A. Kumar, L.D. Stephenson and J.N. Murray, *Progress in Organic Coatings*, 2006, **55**, 3, 244.

[152] J.N. Murray, L.D. Stephenson and A. Kumar, *Progress in Organic Coatings*, 2003, **47**, 2, 136.

[153] L.J. Ballin, *Formulation of a Product Containing the Multifunctional Corrosion Inhibitor System DNBM*, Report No. NADC-90049-60, Air

Vehicle and Crew Systems Technology Department, Naval Air Development Center, Warminster, PA, 1989.

[154] V.S. Agarwala and D.W. Beckert, *Electrochemical Impedance Spectroscopy of Trivalent Chromium Pre-treated Aluminum Alloys*, Report No. NAWCADWAR-94014-60, Naval Air Development Center, Warminster PA, 1993.

[155] F. Pearlstein and V.S. Agarwala, *Proceedings of the AESF Conference on the Search for Environmentally Safer Deposition Processes for Electronics*, Orlando, FL, 1993, p. 112.

[156] N.K. Mehta and M.N. Bogere, *Progress in Organic Coatings*, 2009, **64**, 4, 419.

[157] ASTM D5894, *Practice for Cyclic Salt Fog/UV Exposure of Painted Metal (Alternating Exposures in a Fog/Dry Cabinet and a UV/Condensation Cabinet)*, 2011.

Chapter 5

Self-healing evaluation techniques

Various methods are used for the evaluation of the healing efficiency. One of the three fracture modes (i.e., Mode I, II or III, Figure 5.1) is induced in two sets of devices. The first set of devices includes samples from the original host material; the second set includes samples containing the self-healing agent and the catalyst. After the healing process is completed, a standard test is performed to compare the two sets of the devices. A second test can be run in parallel, or separately, to validate the results from the first test.

Some of the common tests used to measure self-healing efficiency are as follows:

- Stretching the sample up to its rupture with a material testing system (MTS) instrument. The MTS provides the measurement of the extension or displacement as a function of the pulling force. The sample is built in a tapered double-cantilever beam (TDCB). It is slowly stretched up to its failure (damage). This test takes a few minutes to complete. It induces the necessary damage that allows the evaluation of the efficiency of the self-healing process [1–3].
- Three and four-point flexure bend tests where the sample is pressed up to its failure. Some other types of MTS instruments are used that measure the pressing force and the corresponding displacement. The test takes a few minutes to complete. It is used to induce the damage within samples made from carbon fibre reinforced polymer(s) (CFRP) laminates with embedded microcapsules or hollow fibres containing the healing agent [4]. Three or four-point bending test can be followed by a compression after impact (CAI) test [4]. The CAI test is applied to qualify components for aerospace applications. The four-point bend test may be used in combination with an acoustic emission sensor to measure the healing efficiency. The sensor may also detect hidden cracks for systems with a microvascular reservoir [5].
- Indentation test. This test is similar to the three-point bend test set up with a conical shaped mass that is dropped from a certain height [6]. The weight and speed of the dropping mass can be monitored to reproduce a partial or a complete rupture.
- Ballistic test with a projectile. The speed of a projectile is between 300 m/s and 900 m/s [7,8].

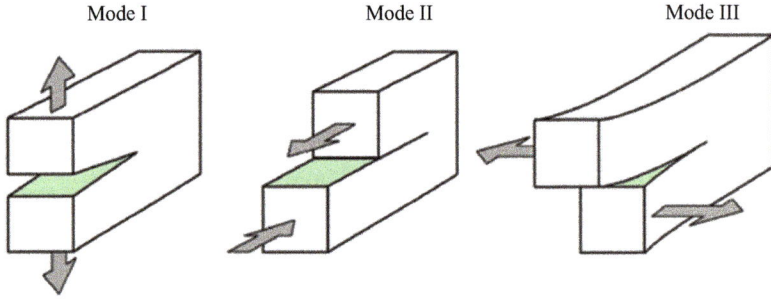

*Figure 5.1   Three basic fracture modes. Mode I: opening (or antiopening)
under tensile (or compressive) stress; Mode II: plane shear
(shear in direction perpendicular to crack front) and Mode III:
anti-plane shear (shear in direction parallel to crack front)*

- Hypervelocity impact test. In this test pellets at a speed between 1 and
  20 km/s are used. The pellets induce craters or holes, similar to those
  made by space debris. This test was applied to the test of CFRP laminates embedding self-healing agent and fibre sensors [9].
- Damage induced by accelerated ageing such as thermal shocks occurred
  by the exposure to an extreme thermal gradient, such as between liquid
  nitrogen (-196 °C) and room temperature [3]. In this test randomly
  shaped cracks are created, which are similar to those naturally developed
  with ageing (Figure 5.2).

5.1   Methods with a three- and four-point bend test

This approach has been applied when the material host is formed from
CFRP laminate layers [9–11]. The samples prepared were small rectangular
shapes, with dimensions which comply with the corresponding American
Society for Testing and Materials (ASTM) test. For example, Aïssa *et al.* [9]
prepared samples with four CFRP layers with a span length of 80 mm,
width of 12 mm and thickness of 5 mm in accordance with the ASTM D790
[12]. In some of the samples, fibre Bragg gratings (FBG) sensors were
embedded. The sensors were used to measure the strain or the applied force.
The MTS instrument provided curves of displacement as a function of the
applied force. The strain measured with the fibre sensors was directly proportional to the applied force. Moreover, the fibre sensors permitted monitoring of small changes in the residual strain over long periods of time
(e.g., from several months to 1 year). The sensors also allowed recording of
the occurring then the evolution with time of the healed material in nondestructive evaluations. Figure 5.3 shows the MTS instrument with a strip

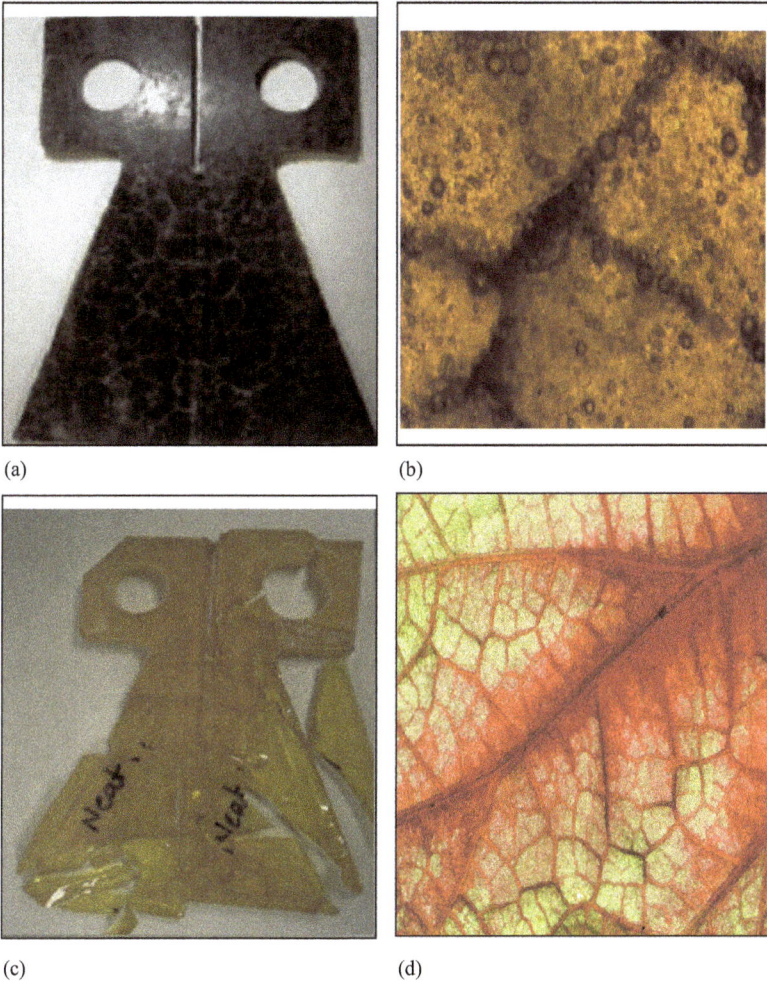

Figure 5.2 *(a) TDCB sample with healing agent after 20 cycles of thermal shock; (b) detail of the healed cracks; (c) sample without healing agent broken after the first thermal shock and (d) picture of a leaf [6] showing the similarity between the TDCB cracks and the fractals in the leaf*

of four layers of woven of CFRP with self-healing materials and fibres sensors embedding in them (details of the fabrication process are described in Chapter 7).

Figure 5.4 shows the regions of different regimes during a three-point bending test. Region I is the elastic condition with a linear variation of displacement as a function of the applied force. Region II is the inelastic

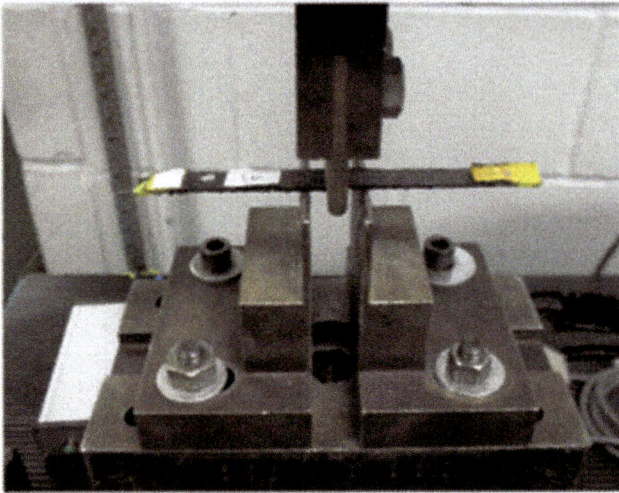

Figure 5.3   *MTS instrument used for the three-point flexure test of CFRP samples with self-healing agent. The distance between the two limits is 2 cm, and the hitting mass is in the middle for the samples of four layers of woven laminates (in accordance with ASTM D790 standard [12])*

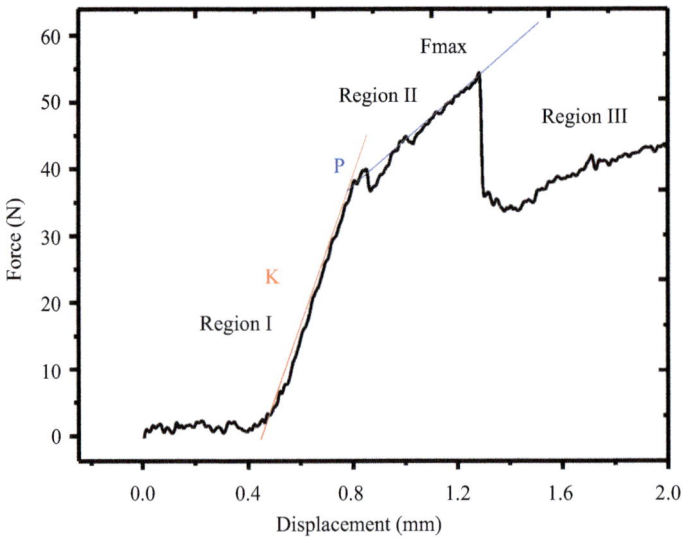

Figure 5.4   *Measurements obtained with the three-point bending test. Region I corresponds to the elastic domain, followed by an inelastic region, delamination and cracks develop in region III. K is the slope of region I. P is the force reached at the interface between regions I and II, and F_{max} is the maximum force reached before starting the failure regime*

condition, which is generally described by a nonlinear displacement in relation to the applied force. Region III is the condition after failure.

The following equations are used to evaluate the mechanical properties (they are deduced from ASTM D790 [12]).

Flexural stress:

$$\sigma_f = 3P_{max}L/2bd^2 \tag{5.1}$$

where σ_f is the stress in the outer fibres at mid-point, in MPa; P_{max} is the maximum load (force) in the elastic regime, in N; L is the support span (mm); b is the width of beam tested (mm) and d is the depth of beam tested (mm).

Modulus of elasticity:

$$E_B = mL^3/4bd^3 \tag{5.2}$$

where E_B is the modulus of elasticity in bending (MPa), L is the support span (mm) and m is the slope of the tangent to the initial straight-line portion of the load-deflection curve (N/mm of deflection).

Shear strength:

$$T = 0.75 \times (F_{max}/b \times h) \tag{5.3}$$

P point strength:

$$T_P = 0.75 \times (P_{max}/b \times h) \tag{5.4}$$

Shear modulus:

$$E = K/b \times h \tag{5.5}$$

where h is the thickness of the beam, in mm and K is the slope of the elastic regime of Region I, in N/mm.

5.2 Tapered double-cantilever beam

Patel *et al.* used a TDCB for the self-healing test [7]. The shape and dimensions (Figure 5.5) were proposed in the 1960s by Mostovoy *et al.* [13], and later used by Beres *et al.* [10]. The MTS instruments used for testing the TDCB provided curves of displacement as a function of the applied force, similar to the three- and four-point bend test (see Figure 5.6). To measure the healing effect, a cut is made along the axis and the sample is left for a certain time, to ensure the healing process is completed. Then the performance of the original samples and those containing a healing agent is compared. The purpose of using the TDCB shape and dimensions proposed

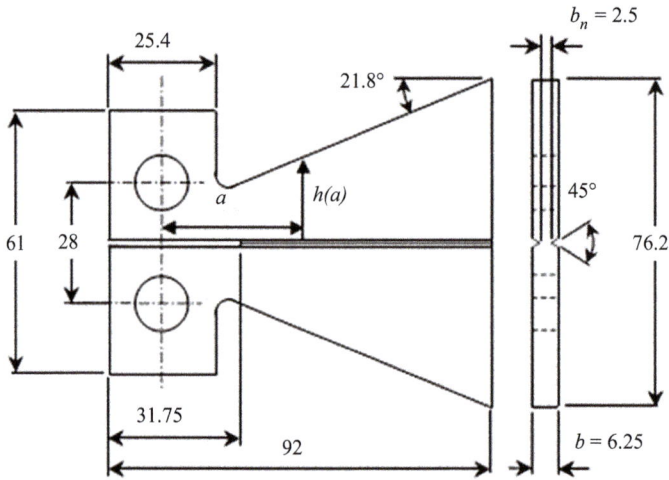

Figure 5.5   *TDCB geometry (dimensions are in mm): a, b and b$_n$ are the crack length, the sample width and the crack width, respectively*

Figure 5.6   *MTS-10 instrument used for stretching the TDCB sample*

by Mostovoy *et al.* [13] is to quickly obtain the results, which are independent of the length of the cut.

The healing efficiency (η) can be determined by measuring the fracture toughness, which is independent of the crack length (5.6):

$$K_{\text{IC}} = \alpha P_C \tag{5.6}$$

where K_{IC} is the fracture toughness; P_C is the critical fracture load and α is the geometric term determined experimentally.

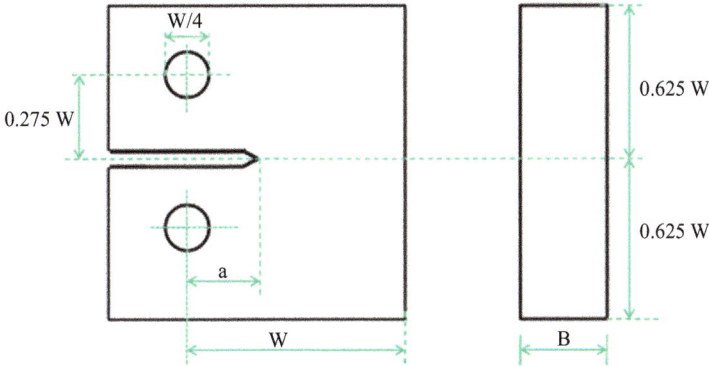

Figure 5.7 Compact tension configuration according to ASTM D5045-99 [14] (dimensions given in inches)

The healing efficiency is defined as

$$\eta = K_{\text{IC-healed}}/K_{\text{IC-virgin}} \tag{5.7}$$

For the TDCB sample geometry, the healing efficiency is given as

$$\eta = \left(K_{\text{IC-healed}}/K_{\text{IC-virgin}}\right) = \left(P_{C\text{-healed}}/P_{C\text{-virgin}}\right) \tag{5.8}$$

The healing efficiency can be calculated from the peak loads during fracture testing.

Brown *et al.* [11] investigated the healing efficiency of TDCB geometry (see the MTS-10 instrument used for TDCB stretching in Figure 5.6). They found that the self-healing polymer recovers as much as 90% of its original fracture toughness.

Smaller samples with different shape have been used (Figure 5.7) in accordance with ASTM D5045-99 [8,14].

Many other alternative shapes can be found in the literature [15]. For example, two other shapes are illustrated in Figures 5.8 and 5.9.

5.3 Compression after impact

With the intention of performing a more stringent assessment of the healing effect, CAI was applied to measure the residual compressive strength of CFRP after damage. The CAI provides a rigorous assessment of a material's performance following a low-velocity impact event. The material's compressive strength is extremely sensitive to internal damage and provides a critical evaluation of self-healing efficiency. The CAI was applied in a test of healing efficiency [4] in conjunction with ASTM D7137/D7137M-05 [14] standard.

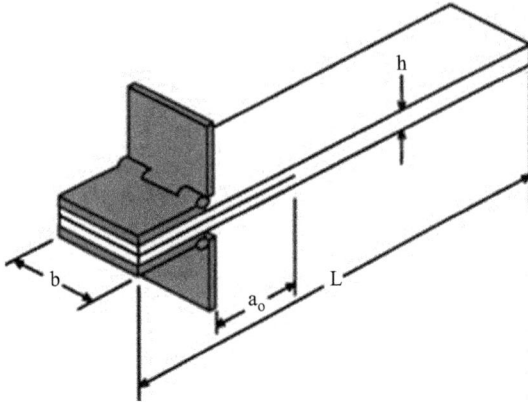

Figure 5.8   Double cantilever beam specimen
[Reproduced, with permission, from [15], © 2001 Elsevier]

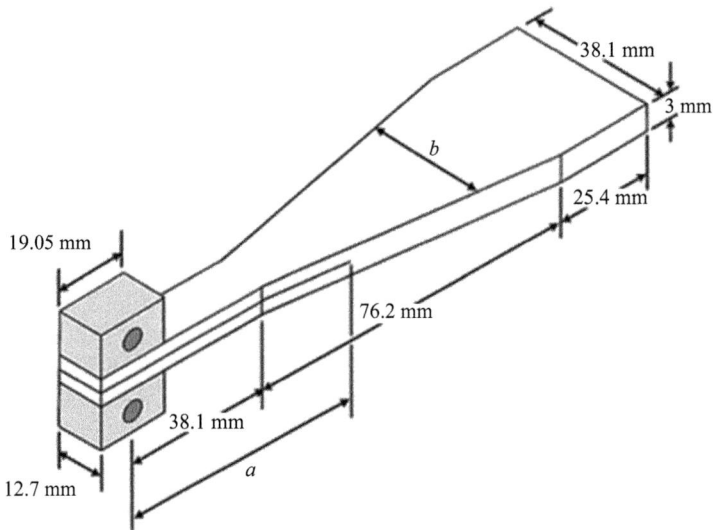

Figure 5.9   Geometry of the wide TDCB specimen used in the fracture
experiments
[Reproduced, with permission, from [15], © 2001 Elsevier]

5.4   Combining the four-point bend test and acoustic emission

Acoustic emission may be used in a nondestructive test. This approach is applied to confirm the healing efficiency data that were determined by another method. Toohey *et al.* [5] used an acoustic emission sensor

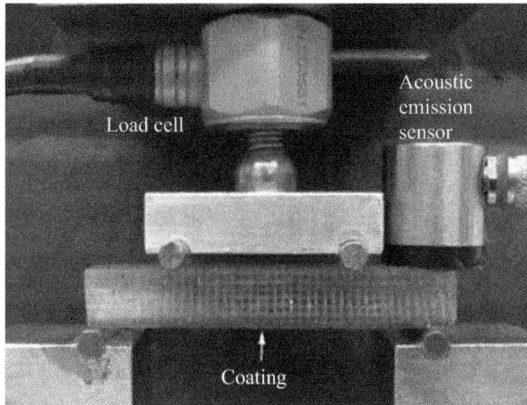

Figure 5.10   Four-point bending test with an acoustic emission sensor
fixed to the specimen
[Reproduced, with permission, from [5], © 2009 Springer]

(Figure 5.10) to validate their experiment. Their test showed different responses of the acoustic emission sensor to the well healed and the poorly healed specimens.

5.5   Methods with dynamic impact

5.5.1   Indentation test with a dropping mass

The set-up is similar to the three-point bend test. In a standard three-point bend test, the MTS instrument will apply an increasing force and measures the deflection in the sample. The partial rupture caused by the MTS instrument is not easily reproducible. With the dropping mass, the partial rupture can be controlled by the hitting weight, the height of the dropping mass and the size of the tip (Figures 5.10 and 5.11). The hitting energy (E) and velocity (v) are deduced from the height (h), the mass (m) and the gravity (g).

Yin *et al.* [16], Patel *et al.* [17] and Yuan *et al.* [18] investigated the microcapsule-based self-healing of composites under low-velocity impact (Figure 5.12). Yin *et al.* [16] investigated an epoxy-loaded microcapsule system for the self-healing of glass/epoxy composite. Their healing agent was used for repairing matrix cracks under low-velocity impact. However, high temperature and pressure were required for the system to achieve good healing. Yuan *et al.* [18] investigated epoxy and mercaptan-loaded microcapsules for the self-healing of the glass/epoxy system and found it to work in an autonomic way at room temperature.

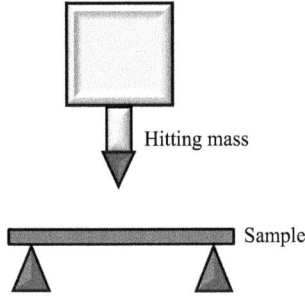

Figure 5.11   Schematic of a three-point bending test with a dropping mass tower

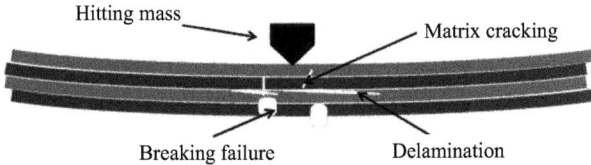

Figure 5.12   Schematic demonstrating the principal damage mechanisms in composite laminates for low-velocity impact

5.5.2   *High-speed ballistic projectile*

Patel *et al.* proposed self-healing of bullet induced damages [7]. The ballistic projectile with a speed of a few hundred m/s (usually between 200 and 900 m/s) is used as a representative of a rifle bullet. When a bullet makes a hole, the material of the hole is lost. In this case the surface tension and gravity prevent the possibility to fill that hole (i.e., its healing). Asgar *et al.* [8] presented a preliminary approach to self-healing of the composite after impact with small pellet launched at approximately 600 m/s.

Patel *et al.* [17] studied the low-velocity impact performance of glass/epoxy composites with urea-formaldehyde microcapsules containing dicyclopentadiene and Grubbs catalyst contained in paraffin wax microspheres. Their study revealed a significant recovery of the residual compressive strength in the composite panels. Patel *et al.* [7] also investigated the same self-healing system with composite armour. They demonstrated the significant recovery of strength for the low-velocity impact and found that the modes of damage in ballistic-impacted composites (up to 475 m/s) are very similar to those of the low-velocity impact.

5.5.3 *Hypervelocity impact*

Hypervelocity impacts (>1 km/s and up to 25 km/s) correspond to debris and micrometeorites impacts on satellites, particularly those at low earth orbital orbits [9]. Their impacts are different from the high velocities. The more detailed data on self-healing of impacts caused by space debris are presented in Chapter 7.

5.6 Fibre Bragg grating sensors for self-healing detection

Sensors based on FBG have attracted considerable attention since the early days of their discovery. FBG sensors exceed other conventional electric sensors in many aspects, for example, they have immunity to electro-magnetic interference, compact size, light weight, flexibility, stability, high-temperature tolerance, and resistance to harsh environments. Additional advantages of FBG sensors include linearity in response over many orders of magnitude. In addition, the embedded FBG sensor systems may be implemented in the very early fabrication stages and then used continuously until the end of life of a given device. Therefore, the parameters describing the structure can be recorded during its whole life cycle permitting better analysis of the evolution of fatigue and microcracks.

Within the specific self-healing process, the fibre sensors provide two levels of information: (i) a dynamic level detecting the impact up to hypervelocity level, with a commercially available system up to 2 MHz [9], and (ii) a static level which compares the residual stress–strain level before and after the impact events.

The optical fibres are cylindrical silica waveguides. They consist of a core surrounded by a concentric cladding, with different refraction index guaranteeing the light propagation (Figure 5.13(a)). FBG (Figure 5.13(b)) are formed when a periodic variation of the refraction index of the core is created along a section of an optical fibre, by exposing the optical fibre to an interference pattern of intense ultraviolet light [19]. If the broadband light is travelling through an optical fibre containing such a periodic structure, its diffractive properties cause a preselected wavelength band to be reflected back (Figure 5.14). The centre wavelength, λ_B, of this band can be represented by the well-known Bragg condition (5.9):

$$\lambda_B = 2n_0 \times \Lambda_B \tag{5.9}$$

where Λ_B is the spacing between grating periods, and n_0 is the effective index of the core.

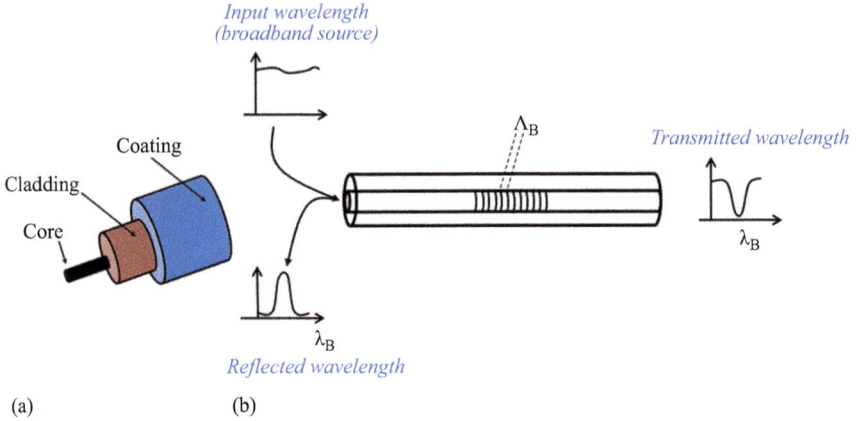

Figure 5.13   (a) Schematic of the optical fibre and (b) light phenomena in the FBG sensor

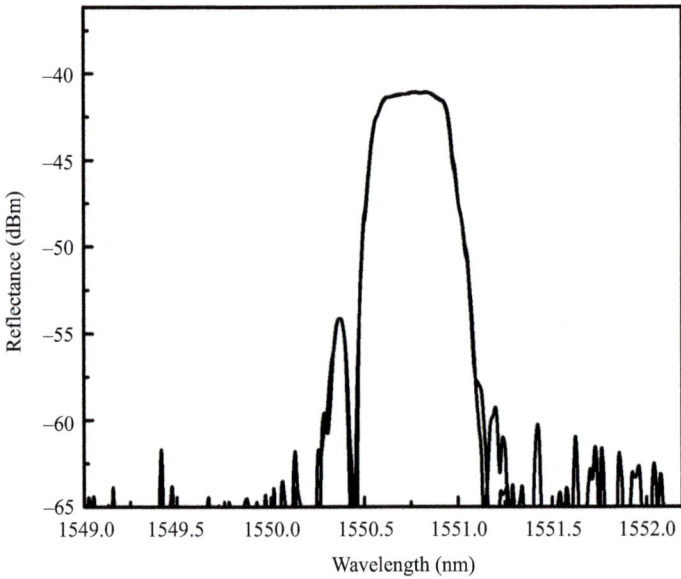

Figure 5.14   Typical optical power spectrum (reflectance in dBm) of the FBG

Gratings are simple, intrinsic sensing elements and give an absolute measurement of the physical perturbation to which they are sensitive. Their basic principle of operation is to monitor the wavelength shift associated with the Bragg resonance condition. The wavelength shift is independent of the light source intensity. At constant temperature $\Delta\varepsilon$, the corresponding wavelength shift is given by (5.10):

$$\Delta\lambda_B = \lambda_B(1/\Lambda_B \times \partial\Lambda_B/\partial\varepsilon + 1/n_0 \times \partial n_0/\partial\varepsilon)\Delta\varepsilon = \lambda_B(1 - p_e)\Delta\varepsilon \quad (5.10)$$

where p_e is the effective photoelastic coefficient of the optical fibre.

For example, for a silica fibre, the FBG wavelength–strain sensitivity at 1,550 nm is 1.15 pm $\mu\varepsilon^{-1}$ (picometre by microstrain) [20].

Figure 5.14 shows a representation of the optical spectrum of FBG observed with the optical spectrum analyser.

When an axial stress is applied to the optical fibre, the wavelength of the reflected spectrum shifts. This shift is to higher wavelengths for axial tension and to lower wavelengths for axial compression. The axial strain applied to the optical fibre at the specific location of FBG can be calculated from the shift in the peak wavelength by using (5.11):

$$\varepsilon = \Delta\lambda/\lambda_B(1 - p_e) \quad (5.11)$$

where $\Delta\lambda$ is the peak wavelength shift, λ_B is the Bragg wavelength, p_e is the effective strain-optic coefficient for the optical fibre fundamental mode and ε is the axial strain [21].

When nonuniform axial strains or transverse stress components are applied to the optical fibre, the reflected spectrum of the FBG is no longer a single peak. The reflected spectrum can be broadened. It takes the form of multiple peaks or a more complex spectral form. This spectral distortion is often used to detect the presence of stress. For nonhydrostatic loading cases, the transverse loading creates birefringence in the optical fibre, leading to two axes of propagation in the fibre. The light wave propagating through the fibre is split into two modes, each experiencing a slightly different Bragg wavelength as they pass through the FBG. When recombined, the reflected spectrum demonstrates two distinctive peaks such as shown in Figure 5.15. The wavelength separation between these peaks is proportional to the magnitude of the transverse stress component [22]. Additionally, nonuniform axial strain along the FBG can further distort the response spectrum.

The centre wavelength of light reflected back from a Bragg grating depends on the periodic spacing between the grating planes. It is also affected by changes in temperature and strain. The shift in the Bragg grating wavelength due to temperature and strain change is given by (5.12) [23]:

$$\Delta\lambda_B = 2(\Lambda\partial n/\partial T + n\partial\Lambda/\partial T)\Delta T + 2(\Lambda\partial n/\partial T + n\partial\Lambda/\partial l)\Delta l \quad (5.12)$$

where $\Delta\lambda_B$ is the Bragg wavelength shift, Λ is the periodicity of the grating, n is the index of refraction of the core, ΔT is the fractional wavelength shift for a temperature change and Δl is the shift of the grating length.

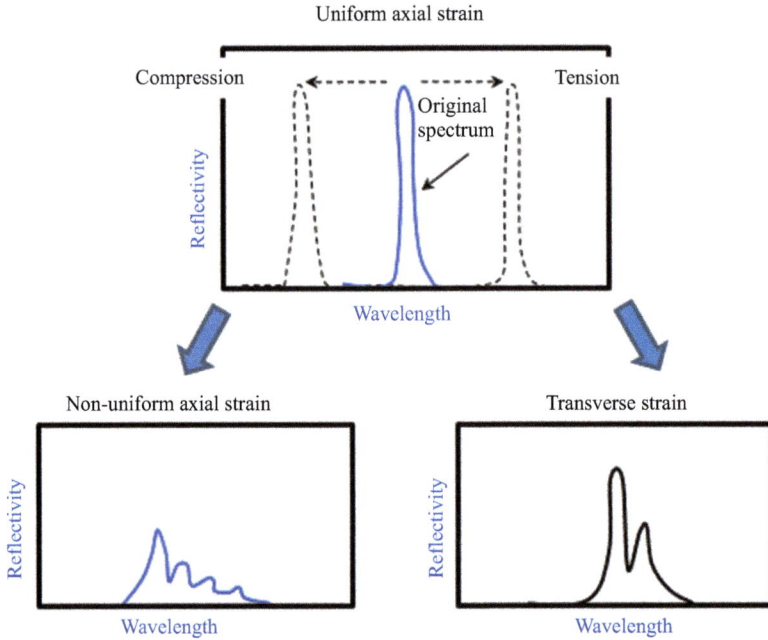

Figure 5.15   Schematic of FBG sensor reflected spectrum under various strain states

The first term represents the temperature effect on an optical fibre. The changes in the grating spacing and the index of refraction caused by thermal expansion result in a shift in the Bragg wavelength. This wavelength shift for a temperature change ΔT may be written as [24]:

$$\Delta\lambda_{B,T} = \lambda_B(\alpha + \xi)\Delta T \tag{5.13}$$

where α is the $(1/\Lambda)(\partial\Lambda/\partial T)$ is the thermal expansion coefficient of the fibre ($\sim 0.55 \times 10^{-6}\,°C^{-1}$ for silica).

The quantity $\xi = (1/n)(\partial n/\partial T)$ represents the thermo-optic coefficient, which is $8.6 \times 10^{-6}\,°C^{-1}$ for a germanium doped silica-core fibre. Clearly, the index change is by far the most dominant effect. From (5.12), the expected temperature sensitivity of the FBG at 1,550 nm is 0.0142 nm/°C.

References

[1]   A. Yavari, K.G. Hockett and S. Sarkani, *International Journal of Fracture*, 2000, **101**, 4, 365.

[2]   M.R. Kessler, N.R. Sottos and S.R. White, *Composites Part A: Applied Science and Manufacturing*, 2003, **34**, 8, 743.

[3]   G. Thatte, S.V. Hoa, P. Merle and E. Haddad, 'Self-healing epoxy for space applications', in *Proceedings of the First International Conference on Self-Healing Materials*, Eds., A.J.M. Schmets and S. van der Zwaag, CD-ROM, 18–20 April 2007, Noordwijkaan Zee, The Netherland, Springer, 2007, Paper No. 12.

[4]   G.J. Williams, I.P. Bond and R.S. Trask, *Composites Part A: Applied Science and Manufacturing*, 2009, **40**, 9, 1399.

[5]   K.S. Toohey, N.R. Sottos and S.R. White, *Experimental Mechanics*, 2009, **49**, 5, 707.

[6]   C. Semprimoschnig, *Enabling Self-healing Capabilities: A Small Step to Bio-mimetic Materials*, Report No. 4476, European Space Agency, Noordwijk, the Netherlands, 2006.

[7]   A.J. Patel, S.R. White, D.M. Baechle and E.D. Wetzel, *Self-healing Composite Armor: Self-healing Composites for Mitigation of Impact Damage in US Army Applications*, Final Report, Contract No. W911NF-06-2-0003, US Army Research Laboratory, Adelphi, MD, 2006.

[8]   M. Asgar-Khan, S. Hoa, B. Aïssa, E. Haddad and A. Higgins, *Proceedings of the Third International Conference on Self-healing*, University of Bristol, Bath, UK, 2011.

[9]   B. Aïssa, K. Tagziria, E. Haddad, *et al.*, 'The self-healing capability of carbon fibre composite structures subjected to hypervelocity impacts simulating orbital space debris', *ISRN Nanomaterials*, 2012, **2012**, Article ID 351205, 16 pages.

[10] W. Beres, A.K. Koul and R. Thamburaj, *Journal of Testing and Evaluation*, 1997, **25**, 6, 536.

[11] E.N. Brown, N.R. Sottos and S.R. White, *Experimental Mechanics*, 2002, **42**, 4, 372.

[12] ASTM D790, *Standard Test Methods for Flexural Properties of Unreinforced and Reinforced Plastics and Electrical Insulating Materials*, 2010.

[13] S. Mostovoy, P. Crosley and E. Ripling, *Journal of Materials*, 1967, **2**, 661.

[14] ASTM D7137/D7137M-05, *Standard Test Method for Compressive Residual Strength Properties of Damaged Polymer Matrix Composite Plates*, 2012.

[15] M.R. Kessler and S.R. White, *Composites Part A: Applied Science and Manufacturing*, 2001, **32**, 5, 683.

[16] T. Yin, M.Z. Rong, J.S. Wu, H.B. Chen and M.Q. Zhang, *Composites Part A: Applied Science and Manufacturing*, 2008, **39**, 9, 1479.

[17] A.J. Patel, N.R. Sottos, E.D. Wetzel and S.R. White, *Composites Part A: Applied Science and Manufacturing*, 2010, **41**, 3, 360.

[18] Y.C. Yuan, Y. Ye, M.Z. Rong, *et al.*, *Smart Materials and Structures*, 2011, **20**, 1, 015024.

[19] R. de Oliveira, C.A. Ramos and A.T. Marques, *Computers and Structures*, 2008, **86**, 3–5, 340.

[20] W.W. Morey, G. Meltz and W.H. Glenn, *Fiber Optic and Laser Sensors VII*, Eds., R.P. DePaula and E. Udd, *SPIE Conference Volume 1169*, Bellingham, WA, 1989.

[21] E. Kirkby, R. de Oliveira, V. Michaud and J.A. Månson, *Composite Structures*, 2011, **94**, 1, 8.

[22] A. Panopoulou, T. Loutas, D. Roulias, S. Fransen and V. Kostopoulos, *Acta Astronautica*, 2011, **69**, 7–8, 445.

[23] A. Othonos and K. Kalli, *Fiber Bragg Gratings: Fundamentals and Applications in Telecommunications and Sensing*, Artech House Publishers, Norwood, MA, 1999.

[24] G. Meltz and W.W. Morey, *International Workshop on Photoinduced Self-organization Effects in Optical Fiber*, Ed., F. Ouellette, *SPIE Conference Volume 1516*, Bellingham, WA, 1991.

Chapter 6

Review of advanced fabrication processes

In this chapter, the main experimental results obtained to date on the self-healing composite materials are reviewed. The review starts with the nanostructuration of the ruthenium Grubbs' catalyst (RGC) by means of the laser ablation process, followed by the encapsulation of the 5-ethylidene-2-norbornene (ENB) liquid monomer into small capsules and the fabrication of three-dimensional (3D) microvascular nanocomposite beams by microfluidic infiltration. Special attention is given to the use of single-wall carbon nanotubes (SWCNT) material as reinforcement of the ENB healing agent from the perspective of obtaining a self-healing composite material with improved mechanical properties and, at the same time, having a fast ring-opening metathesis polymerisation (ROMP) reaction with high mechanical properties.

6.1 Ruthenium Grubbs' catalyst

In this section, the preparation of a self-healing composite material that consists of ENB monomer reacted with RGC is reviewed. First, the kinetics of the ENB ROMP reaction with RGC was studied as a function of temperature. It was shown that the polymerisation reaction was still effective over a large temperature range ($-15\,^{\circ}$C to $45\,^{\circ}$C), occurring at short time scales (less than 1 minute at $40\,^{\circ}$C). Second, the amount of RGC required for the ROMP reaction was found to decrease significantly through its nanostructuration by means of an ultraviolet (UV)-excimer laser ablation process. RGC nanostructures of a few nanometers in size were obtained directly on silicon substrates. The X-ray photoelectron spectroscopy (XPS) data strongly suggest that the RGC still keeps its original stoichiometry after nanostructuration. More importantly, the associated ROMP reaction was successfully achieved at an extreme low RGC concentration equivalent to $11.16 \pm 1.28 \times 10^{-4}$ vol%, occurring in a very short time reaction [1]. This approach opens new prospects for using nanocomposite materials as healing agents for self-repair functionality, thus obtaining a higher catalytic efficiency per unit mass.

6.1.1   Pulsed laser deposition technique

Pulsed laser deposition (PLD) is a versatile technique. In this method the energy source is located outside the chamber, therefore the use of a ultra-high vacuum (UHV) as well as ambient gas is possible. Combined with a stoichiometry transfer between the target and substrate, this allows deposition of all kinds of materials, for example, high-temperature super-conductors, oxides, nitrides, carbides, semiconductors or metals. Even polymers or fullerenes can be grown with high deposition rates [2]. The nature of the PLD process allows the preparation of complex polymer-metal compounds and multilayer films. In UHV, implantation and intermixing effects originating in the deposition of energetic particles lead to the for-mation of metastable phases, for example, nanocrystalline highly super-saturated solid solutions and amorphous alloys. The preparation in an inert gas atmosphere makes it possible to control the film properties (stress, texture, reflectivity, magnetic properties and so on) by varying the kinetic energy of the deposited particles. All this makes PLD an attractive deposi-tion technique for the growth of high quality, thin films and nanostructures.

With the PLD method, thin films are prepared by the ablation of one or more targets illuminated by a focused pulsed-laser beam. This technique was first used by Smith and Turner [3] in 1965 for the preparation of semiconductors and dielectric thin films. In 1987 it was then fully devel-oped by Dijkkamp *et al.* [4] for the deposition of high-temperature super-conductors. Their work allowed them to define the main characteristics of PLD, namely the stoichiometry transfer between the target and the depos-ited film, high deposition rates of about 0.1 nm per pulse and the occurrence of droplets on the substrate surface [2–5]. Since the work of Dijkkamp *et al.*, the deposition technique has been used extensively for all kinds of oxides, nitrides and carbides and also for preparing metallic systems and even polymers or fullerenes. During PLD, many experimental parameters can be modified, which then have a strong influence on the properties of the deposited film. First, the laser parameters such as laser fluence, wavelength, pulse-duration and repetition rate can be adjusted. Second, the deposition conditions, including target-to-substrate distance, substrate temperature, background gas and pressure may be adjusted to control the film growth.

A typical set-up for PLD is shown schematically in Figure 6.1. In a UHV chamber, elementary or alloy targets are struck at an angle of 45° by a focused pulsed laser beam. The atoms and ions ablated from the target(s) are deposited on a substrate. In most cases, the substrate is attached with its surface parallel to the target surface at a distance of 2–10 cm.

Table 6.1 shows a list of materials deposited since the introduction of PLD in 1987. In order to obtain all these different kinds of materials, one

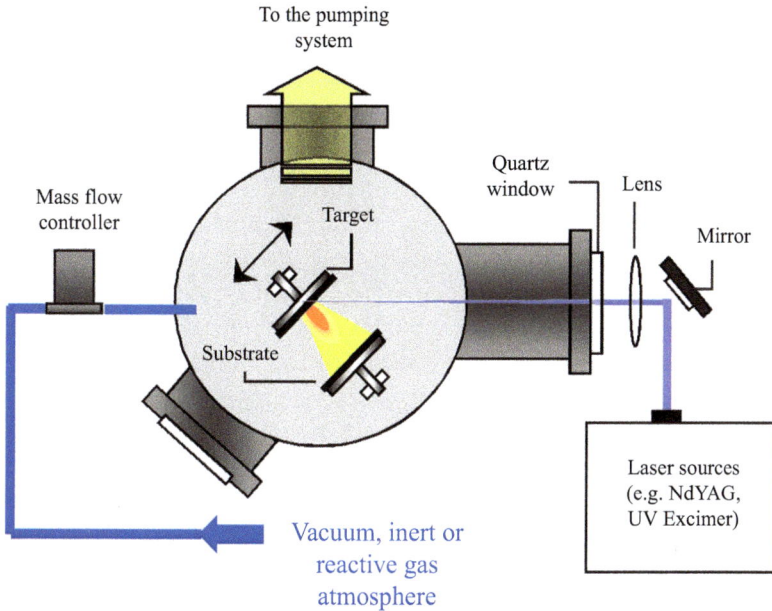

Figure 6.1 Schematic diagram of a typical laser deposition set-up neodymium-doped yttrium aluminium garnet (NdYAG)

Table 6.1 List of some materials deposited for the first time by PLD after 1987

Material	Reference
High-temperature $YBa_2Cu_3O_7$	Dijkkamp *et al.* [4]
Superconductors BiSrCaCuO	Guarnieri *et al.* [6]
TlBaCaCuO	Foster *et al.* [7]
MgB_2	Shinde *et al.* [8]
Oxides, for example, SiO_2	Fogarassy *et al.* [9]
Carbides, for example, SiC	Balooch *et al.* [10]
Nitrides, for example, TiN	Biunno *et al.* [11]
Ferroelectric materials	Kidoh *et al.* [12]
Diamond-like carbon (C)	Martin *et al.* [13]
Buckminster fullerene (C_{60})	Smalley and Curl (1991) [14]
Polymers polyethylene, polymethyl methacrylate	Hansen and Robitaille (1988) [15]
Metallic systems: 30 alloys/multilayers	Krebs and Bremert (1993) [16]
FeNdB	Geurtsen *et al.* [17]
Multiferroic Bi_2FeCrO_6	Nechache *et al.* [18]
Nanostructured RGC	Aïssa *et al.* [1]

has to work in UHV or a reactive gas atmosphere. During the growth of oxides, the use of oxygen is often needed to enhance the amount of oxygen in the growing film. For example, for the formation of perovskite structures at high substrate temperatures in a one-step process, an oxygen pressure of about 30 Pa is necessary [4]. Also, for many other oxides or nitride films, the necessity of working in a reactive environment makes it difficult to use other types of deposition techniques, such as thermal evaporation by using electron guns. In the case of sputtering, where argon is commonly used as the background gas, a larger amount of oxygen or nitrogen can only be added if special oven facilities are added close to the substrate surface.

In many cases, one takes advantage of the fact that during PLD, the stoichiometry of the deposited material is very close to the target. Therefore, it is possible to prepare stoichiometric thin films from a single alloy bulk target. This so-called 'stoichiometry transfer' between target and substrate has made the PLD technique even more attractive. For example, it allows the growth of complex systems such as high-temperature super-conductors, piezoelectric and ferroelectric materials with perovskite structure. It is also possible to make devices such as sensors and capacitors. Stoichiometry transfer between target and substrate is difficult to obtain with evaporation or (magnetron) sputtering by using a single target, because the partial vapour pressures and sputtering yields of the components are different from each other, which gives rise to a different concentration of the thin film growing on the substrate. In the case of PLD, a stoichiometry transfer between the target and the substrate is obtained. The stoichiometry transfer may be realised in the following way. First, the fast heating of the target surface is achieved through an intense laser beam impact. Typically, temperatures up to or above 4727 °C within a few ns [3,19] are generated. This corresponds to a heating rate of about 739 °C/s. This high heating rate ensures that all target components, irrespective of their partial binding energies, evaporate at the same time. Such an ablation rate is sufficiently high with the laser fluences that are well above the ablation threshold. A so-called Knudsen layer is formed [20] and a high-temperature plasma [21], which then adiabatically expands in a direction perpendicular to the target surface. Therefore, during PLD, the material transfer between target and substrate can be efficiently implemented.

6.1.2   *Experimental preparation of a ruthenium Grubbs' catalyst-pulsed laser deposition target*

Ruthenium For self-healing applications, the catalyst chosen was a first generation Grubbs' catalyst: *bis*(tricyclohexylphosphine)benzylidene ruthenium(IV)dichloride. Grubbs' catalysts are well known for promoting olefin metathesis, showing high activity while being tolerant of a wide range of functional groups [22]. Dissolution of RGC by the healing agent is a critical parameter for an efficient and rapid ROMP reaction. Previous studies [23] demonstrated that smaller and finer RGC particles uniformly distributed in the polymer matrix result in better dissolution kinetics of the particles. The incorporation of the commercially available RGC particles into the polymer matrix is usually a delicate operation since the particles tend to agglomerate. Besides, RGC materials have high costs rendering them limited for large-scale commercial applications. Reducing the size and amount of the RGC needed by altering the material's nanostructure increases the reactivity per unit mass, thereby improving performance and simultaneously reducing costs.

In the present section an experimental approach based on plasma laser ablation for nanostructuring RGC is described. This approach allows for effective integration of the nanocatalyst in composite structures for self-healing applications. Specifically, a self-healing composite material consisting of an ENB monomer was prepared and then reacted with RGC. All the chemicals (Grubbs' catalyst first generation, ENB and dicyclopentadiene (DCPD) monomers and so on) were purchased from Sigma–Aldrich. The experimental approach for RGC nanostructuring is based on laser-plasma methods (PLD). A solid RGC target (~2.5 cm diameter) was fabricated from the as-received commercial RGC powder by high-pressure sintering. Then, RGC-nanoparticles (NP) were generated by ablating the solid RGC target by means of a UV-KrF-excimer laser ($\lambda = 248$ nm; $t \approx 15$ ns; repetition rates 5–10 Hz, at an incident angle of $45°$). The RGC-NP deposition was achieved by ablating the solid RGC target with an on-target laser fluence of 0.1 J/cm^2 at a nonfocal deposition regime, under an inert helium gas background atmosphere of 1.33 Pa, at room temperature (RT). A silicon substrate (500 µm-thick Si (100)) was located 50 mm away from the target, which has a dual rotational and translational motion to ensure a uniform erosion pattern over the entire target surface. These conditions were selected to yield RGC-NP with a size of a few nanometers, and the density varied, depending on the number of laser pulses. NPs were synthesised at deposition rate of 1,000 laser pulses. RGC-NP specimens were characterised by using contact mode atomic force microscopy (AFM) (NanoScope III, Digital Instrument) operated at RT in ambient air.

6.1.3 *Experimental results*

Ruthenium To study the kinetics of the ROMP reaction, the as-received RGC powder (~1 wt%) was slowly added to the ENB stirred solution (Barnstead hotplate stirrer, model SP131825, Barnstead International).

The composite samples obtained were then placed in a Tenney Junior Environment Chamber^{TM} under air (accuracy $\pm$ 3 °C). The ROMP reactions were then performed at different temperatures. For comparison purposes, the ROMP reactions of the ENB (reacted with RGC-NP) and that of the DCPD monomer (reacted with the RGC powder) were also investigated under the same conditions.

The microscopic optical characterisation was performed under a transmission light optical microscope using the 40× objective (BX61, Olympus) and image analysis software (Image-Pro Plus, Media Cybernetics Inc.). Scanning electron microscopy (SEM) micrographs were obtained by using a Jeol JSM-6300F microscope. The elemental chemical bonding of the samples was investigated by XPS, performed at RT and at a base pressure of 10^{-7} Pa, with the VG ESCALAB 220i-XL system (VG Thermo), using monochromatic AlK$_\alpha$ radiation as the excitation source (1,486.6 eV, full width at half-maximum of the Ag 3d 5/2 line = 1 eV at 20 eV pass energy). The reported binding energies were calibrated with respect to the C 1s line at 284.5 eV. The crystal orientation of the RGC-NP was evaluated by using a θ–2θ X-ray diffractometer [Scintag Inc., (now Thermo Optek Corp.), Cupertino, CA, USA] with CuKα radiation generated at 40 kV and 30 mA.

Figure 6.2 shows typical optical and SEM micrographs of the as-received Grubbs' catalyst showing rods up to ~120 μm in length and ~30 μm in width.

Figure 6.3(a) shows a typical AFM image of the PLD-deposited RGC-NP on silicon substrates. The AFM image shows that the RGC-NP are uniformly distributed on the silicon surface, with a surface density of about 1.6×10^9 cm^{-2} (which is estimated by calculating the number of the deposited particles per unit area). The mean particle size of these RGC-NP was determined from their topographic heights in the AFM images, and their corresponding distribution histogram (fitted with a Gaussian distribution) is shown in Figure 6.3(b). The RGC-NP mean size is found to be 20 ± 8 nm, and represents a size reduction – achieved by laser ablation – of over three orders of magnitude compared to the dimensions of the as-received RGC.

As the chemical reactivity (i.e., the ROMP) depends on the availability of RGC to be exposed to the monomer, it follows that decreasing the size of the RGC will increase its specific surface area and thus, its surface to volume ratio. This, in turn, leads to higher dissolution kinetics and an efficient ROMP reaction. In addition, the RGC nanostructures could be distributed more homogeneously throughout the composite matrix.

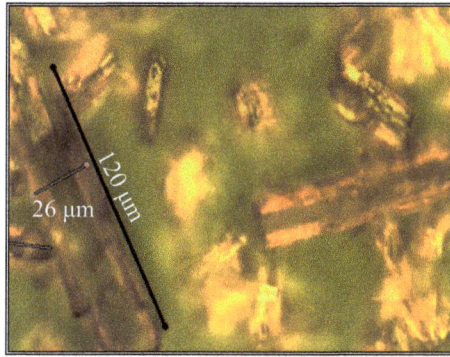

(a)

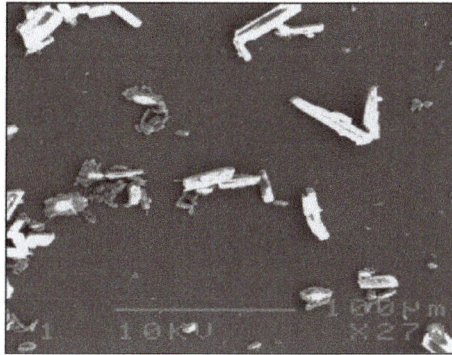

(b)

Figure 6.2   Representative (a) optical and (b) SEM micrographs of the
as-received Grubbs' catalyst showing rods up to ~120 μm in
length and ~30 μm in width
[Reproduced, with permission, from [1], © 2012 Elsevier]

To verify the chemical composition of the generated RGC-NP, the deposits were analysed by XPS. Figure 6.4 shows the typical XPS spectrum of the as-prepared solid RGC target and the PLD RGC-NP. The core level peaks of the main components of the Grubbs' molecules, namely, ruthenium (Ru $3p_{3/2}$ and Ru $3p_{1/2}$ at 463 and 485 eV, respectively), phosphor (P 2p and P 2s located at 132 and 190 eV, respectively) and chlorine (Cl 2p and Cl 2s, located at 200 and 276.5 eV, respectively) are well identified in both the RGC target and NP.

(a)

Aver. size: 20 ± 8 nm

(b)   Diameter (nm)

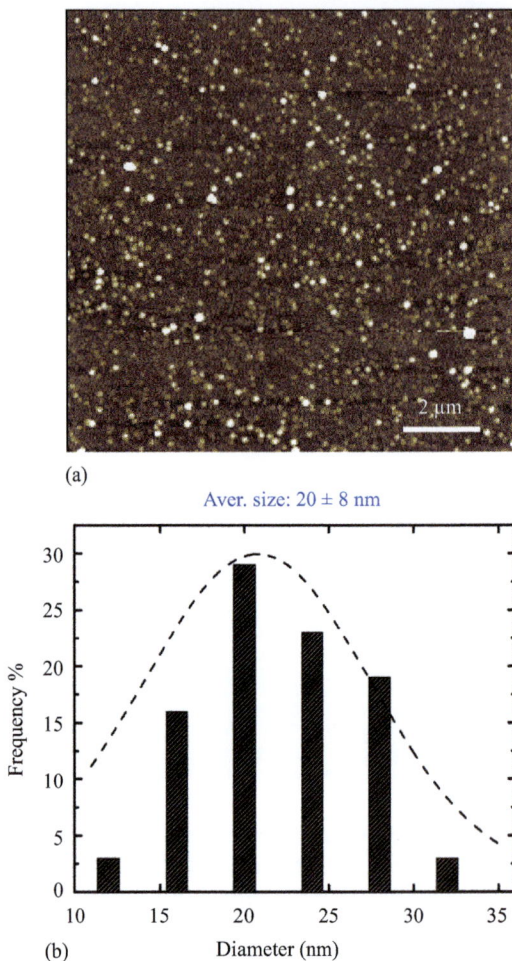

Figure 6.3   *(a) Typical AFM image of the RGC after its laser*
nanostructuration onto a silicon substrate, showing well
distributed NP, having a mean diameter of about 20 ± 8 nm,
as shown by their (b) distribution histogram with Gaussian fit
[Reproduced, with permission, from [1], © 2012 Elsevier]

These results, in addition to the ROMP capability of the RGC-NP,
strongly suggest that the catalyst still keeps its original stoichiometry after
processing by laser ablation. Then, an ENB droplet of a controlled volume
(10 µl) was transferred onto the RGC-NP (on a silicon substrate) using a
micropipette. The sample was kept under ambient conditions of pressure
and temperature, and then observed using SEM after 60 minutes. SEM
observations (Figure 6.5) show that the ROMP reaction has taken place

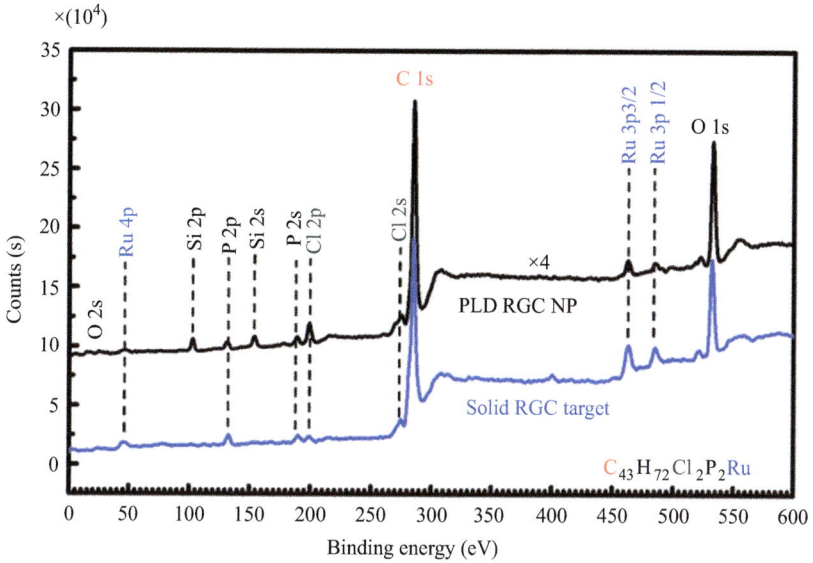

Figure 6.4   XPS pattern of the RGC materials (as-received and PLD nanostructures)
[Reproduced, with permission, from [1], © 2012 Elsevier]

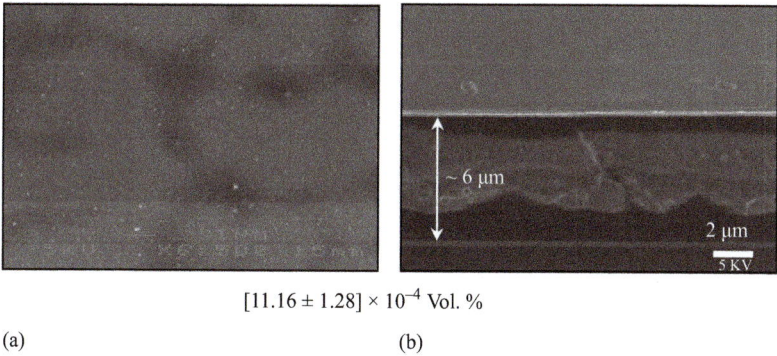

$[11.16 \pm 1.28] \times 10^{-4}$ Vol. %

(a)   (b)

Figure 6.5   Typical SEM micrographs of the ENB reacted with RGC-NP. (a) Top views of the formed polymer film onto silicon substrate and (b) typical cross-sectional SEM view of the ENB polymerised film having a mean thickness of 6 μm
[Reproduced, with permission, from [1], © 2012 Elsevier]

successfully, as homogenous films of ENB/RGC polymer were formed on the silicon substrate. The typical thickness of the polymer formed was around 6 ± 1 μm. Then, by assuming that the RGC-NP needed to this reaction are spherical in shape, and by considering their surface density (which is the number of NP per unit surface area), the ratio of the volume of the RGC-NP relatively to that of the obtained polymer was estimated to be as low as $11.16 \pm 1.28 \times 10^{-4}$ vol%. The concentration is calculated as the ratio of the volume of the NP to the volume of the polymer film formed. Such a catalyst concentration in the ROMP conversion has never been reported yet, as far as is known.

Finally, to point out quantitatively the kinetics of the polymerisation reaction as a function of temperature, ROMP reactions were tested at temperatures ranging from -15 to 45 °C with an RGC load of 1 wt%. Figure 6.6 illustrates the time needed for the ENB polymerisation (ROMP) as a function of reaction temperature. This time is found to be 146 minutes at -15 °C, which decreases to 4 minutes only when the reaction temperature increases to 20 °C. The polymerisation then occurs in a very short time (as low as 0.2 minutes) when the temperature is increased to 40 °C.

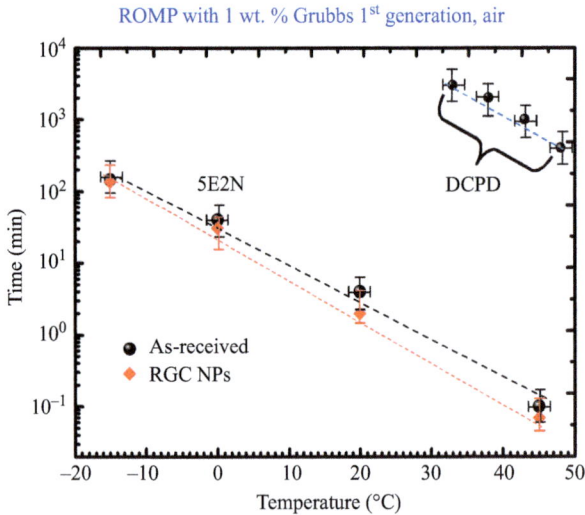

Figure 6.6   Comparison of the ROMP time reaction as a function of reaction temperature, for the ENB polymerisation reacted with both 'as-received' and 'PLD-nanostructured' RGC (first generation Grubbs' catalyst, 1 wt%, in air). The figure includes a comparison with the DCPD healing agent as well [Reproduced, with permission, from [1], ©2012 Elsevier [1]]

The experimental results of the polymerisation time reaction of the DCPD healing agent, reacted with the same load of as-received RGC (i.e., 1 wt%), are included in the same graph for comparison. The polymerisation of the DCPD monomer was found to not only be limited to reaction temperatures starting from 32.5 °C and above [24,25], but their associated reaction times are rather long, ranging from ~400 to 3,000 minutes. It is worth noting as well, that at RT, the DCPD monomer is in its solid state, and is not able to react with RGC [24].

Additionally, the ROMP reaction times associated with RGC-NP are included in Figure 6.6. Table 6.2 shows a summary of the comparison between the ROMP reaction times of the as-received and the PLD-nanostructured RGC materials. This reaction time is found to be the same for the low temperature reaction (~137 minutes at −15 °C), while it decreases to 31 minutes for 0 °C, and is only 2 minutes for a temperature of 20 °C (which represents the half-time value of the 4 minutes needed for the as-received RGC). Finally, the ROMP polymerisation is quasi-instantaneous and is achieved at a time of 0.07 minutes for temperatures of 40 °C. In summary, at RT, the ROMP reaction of the ENB monomer triggered by the Grubbs' catalyst is found to happen in a very short time (less than 4 minutes at RT) for both RGC structures. The ENB monomer definitely provided the best polymerisation time compared to the alternative monomers [26].

The work is in progress to control the size of the RGC-NP with acceptable reproducibility. In fact, understanding the effect of the NP size on the ROMP conversion should help better evaluate their role when integrated in the relevant self-healing systems as well as the physical and chemical mechanism of the self-healing process.

In summary, a successfully synthesised, nanostructured, self-healing composite material consisting of ENB monomer reacted with nanostructures of RGC grown by PLD process. The amount of RGC available and active for the ROMP reaction decreased significantly, to a value of

Table 6.2 Summary of the ROMP reaction times shown in Figure 6.6

Temperature (°C)	Time (min) RGC as-received	Time (min) RGC-NP
−15	146	137
0	40	31
20	4	2
45	0.1	0.07

$11.16 \pm 1.28 \times 10^{-4}$ vol% through its nanostructuration. This approach opens new prospects for using nanocomposite materials containing healing agent for self-repair functionality, thus obtaining a higher catalytic efficiency per unit mass. This work continues with the aim of controlling the reproducibility of the RGC-NP. Second, it aims at integrating these RGC nanostructures in relevant self-healing systems (particularly into microvascular network architecture) in order to study the kinetics of the ROMP polymerisation as a function of both the RGC-NP sizes and the RGC-densities (i.e., concentrations). All the investigations involved have to be systematically corroborated with the appropriate microscopic and spectroscopic observations to clarify the chemical and physical mechanisms of the self-healing process of the healed specimens.

6.2 Healing capability of self-healing composites with embedded hollow fibres

Silica capillary tubes based on hollow fibres were used for this research. They were chosen as a storage reservoir for the healing material. Their internal and outer diameters are summarised in Table 6.3.

6.2.1 Detail of the capillary filling with healing agent

According to the capillary principle, the height of the liquid column in the capillary tube is given by (6.1):

$$h = (2\gamma \cos \theta)/(\rho \cdot g \cdot r) \qquad (6.1)$$

where γ is the liquid-air surface tension (J/m^2 or N/m), θ is the contact angle, ρ is the density of liquid (kg/m^3), g is the acceleration due to gravity (m/s^2) and r is the radius of tube (m).

As an example, the water-filled glass tube at the sea level was considered for calculation purposes to get some idea about the liquid column height for different capillary radii.

For a water-filled glass tube in air at sea level: γ is the 0.0728 J/m^2 at 20 °C, θ is the 20, ρ is the 1,000 kg/m^3 and g is the 9.81 m/s^2.

The height of the water column is given by $h \cong (1.4 \times 10^{-5})/r$.

Using (6.1), the liquid column height in the capillary with different diameters were calculated and are shown in Table 6.4.

The maximum height for the ENB healing agent would be of the same order of magnitude.

Table 6.3   Summary of the hollow fibres (capillary tubes) employed

Set. No	Inner diameter (μm)	Outer diameter (μm)	Optical picture
1	20	90	
2	40	105	
3	50	150	
4	75	150	
5	100	164	

Table 6.4   Maximum height of water column filled in an open-end tube by capillary effect, for different tube diameters

No.	Capillary size (internal diameter, μm)	h (mm)
1	20	1,400
2	40	700
3	50	560
4	75	373
5	100	280

6.2.2   Hollow fibres

Capillaries were cut to 6 mm in length using fibre cleaver. For the proper sealing of capillary ends, the cleaved surfaces were investigated under optical microscopy and found to be flat to the axis of fibre (Figure 6.7(b)). This characteristic has been found to be very different when the capillary was cut by scissors (Figure 6.7(a)).

6.2.3   Capillary filling with ENB healing agent material

Capillaries were filled with ENB monomer. The capillary action is shown in Figure 6.8. ENB monomer acts as healing material and it gets ROMP reaction once contacted with RGC.

After filling, it was necessary to seal the open ends of capillary to ensure that the healing material will remain inside. For sealing, different adhesives/sealants were proposed and tested. Table 6.5 shows the various adhesives/sealants and their effects. Figure 6.9 shows equally spaced fibres

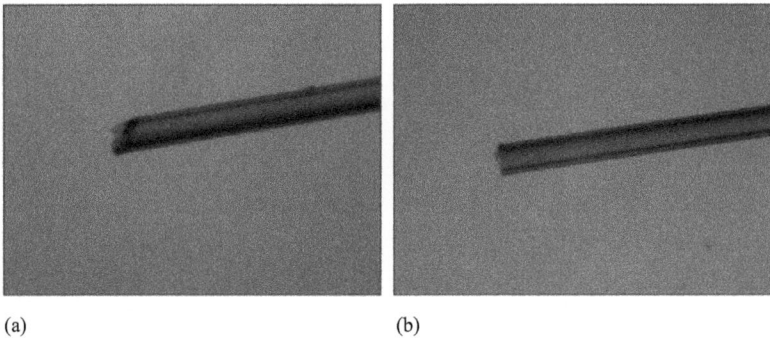

(a)   (b)

*Figure 6.7   Cleaved surface of capillary tubes (a) cut by scissor and
            (b) with standard optical fibre cleaver*

(a)   (b)

*Figure 6.8   (a) Process of the capillary filling with liquid healing agent
            and (b) close up view of the filled capillary end*

Table 6.5 *Properties of various sealants proposed for capping the end of the hollow fibre*

No.	Sealant/adhesive	Curing method
1	Silicone sealant	Non curable
2	LOCTITE (commercial super glue)	Instant room condition curing
3	Epoxy structural adhesive	Needs 24 hours at RT to dry. To achieve full strength, postcuring is needed
4	Durabond 950 FS	Needs 24 hours at RT to dry. To achieve full strength, postcuring is needed
5	Epoxy resin (Epon^{TM} 828 + Epikure^{TM} 3046)	Needs 24 hours at RT to dry. To achieve full strength, postcuring is needed

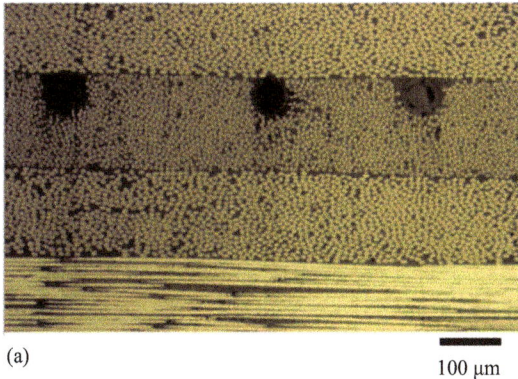

(a)

100 μm

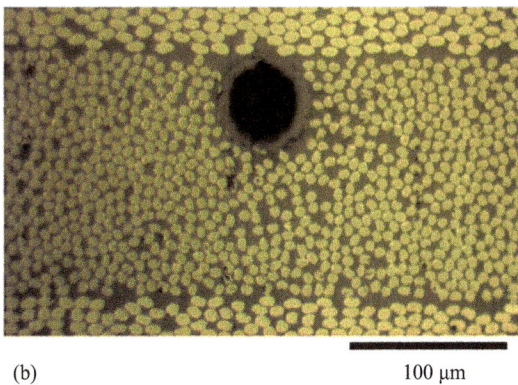

(b)

100 μm

Figure 6.9 *Capillary fibres spaced at 200 μm showing (a) consistent spacing and (b) excellent embedment within host laminate*
[Reproduced, with permission, from [27], © 2007 Royal Society Publishing]

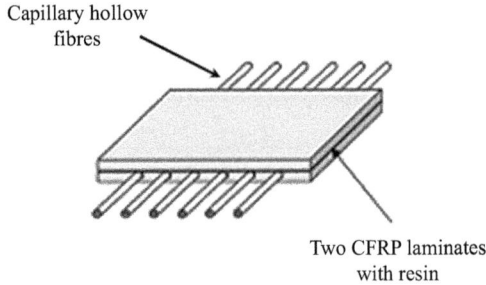

Figure 6.10   Schematic example of embedding the hollow fibres before curing the laminates together

with healing agent embedded between two carbon fibre reinforced polymers (CFRP) layers from [27].

Two sets of hollow fibres were embedded within glass fibre reinforced polymers (GFRP) laminates. The glass permits to observe the healing through the cracks. Figure 6.10 shows a schematic of embedding the hollow fibres before curing the laminates.

6.2.4   Healing with hollow fibres

Hollow fibres with an internal diameter between 20 and 100 μm were filled with the healing agent ENB.

Two approaches were used:

- Capillary effect: because of the small size of the inner diameter, the healing agent liquid is pulled into the capillary tube by the van der Waals effect. In this case the fibres would be filled with the healing agent before being embedded in CFRP.
- Vacuum-assisted capillary effect: where a slight vacuum permits the healing liquid to be aspirated within the hollow fibre. The fibres would be embedded in CFRP before being filled with healing agent in vacuum.

In both cases there is a slight leak at high temperature and long-term hermetic sealing of the capillary ends may require more investigation, for example, use more than one epoxy, that is, where an initial fast curing epoxy seals for a short period, and is assisted by the use of a second epoxy for the long-term (Table 6.6).

Fluorescein ($C_{20}H_{12}O_5$) is commonly used as a dye and its fluorescence is very high. The excitation of the fluorescein molecule occurs at 494 nm (solar light) and the emission at 521 nm (yellow). Fluorescein dye was

Table 6.6   Hollow fibres filled in vacuum assisted capillary action

Parameter	Description
Healing agent	ENB
Viscosity	Very low
Curing temperature	−85 to +146 °C
Capillary filling method	Vacuum assisted capillary action
Sealants (capillary ends)	Epon^{TM} 828 + Epikure^{TM} 3046 + 1 wt% RGC

Figure 6.11   Cross-section of hollow fibres filled with ENB (without catalyst) and fluorescein embedded within GFRP

successfully combined with the healing process. It flows through the cracks to visualise the healing spread between the glass or carbon fibres. It permits the visualisation of the spread of the healing agent after damage to hollow fibres (Figure 6.11).

The cracks were created with a tip of 4.5 mm diameter, and then the sample was cut along a cross-section passing through the damage. The indentation damage could be induced with different forces estimated from the tip weight and the height of dropping (typically between 1 and 2.5 kN).

The damage with the hollow fibres in GFRP/CFRP can be induced in several different ways:

- Indentation with the same drop weight impact tower (may produce large failure such as major matrix cracking);
- Three- or four-point test bend, applying a fixed static force, to produce internal damage and
- The statistical effects of the induced damage by both methods make the healing evaluation harder, since there are unknown variations of the created defects.

Even after the induced damage, the CFRP still has about 74% of its strength. After healing, the strength is 86% (i.e., a 12% increase in strength is recovered from the healing process).

6.3   Encapsulation of the ENB healing agent inside polymelamine-urea-formaldehyde shell

As mentioned previously, because of the relatively high freezing temperature of DCPD, it was necessary to use another monomer which has a wider temperature range. A number of candidates were evaluated. ENB, which is a monomer with a liquid phase from −80 to 148 °C as an alternative, was chosen for further investigation. Liu *et al.* [28] also show that ENB has a faster reaction rate than DCPD. From the cost and toxicity viewpoint, this monomer is more attractive than the others. While ENB has a wider liquid temperature range than DCPD, it was important to investigate its stability and encapsulation.

6.3.1   Stability of ENB in poly-urea-formaldehyde shells

The poly-urea-formaldehyde (PUF) was used as a shell material (see [24,29–35]).

The ENB, resorcinol, urea and all the chemicals used were purchased from Sigma–Aldrich. Ethylene-maleic anhydride (EMA) was used as an organic emulsifier. Aqueous solutions of 1-octanol and formaldehyde (37 wt%) were purchased from Fisher Scientific. Fabrication of the microcapsules follows the procedure outlined in Figure 6.12.

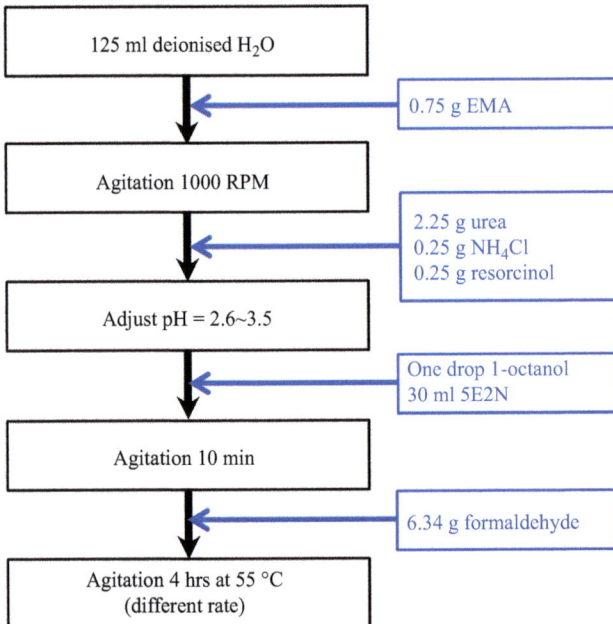

Figure 6.12   Flow chart for process to make microcapsule with PUF shells

Microcapsules were prepared at different stirring rates at 510, 800 and 1,200 rpm. The bulk suspension of microcapsules was filtered and the microcapsules were subsequently washed with deionised water and acetone, then dried in air. The sizes of the microcapsules obtained were measured by optical microscopy, and an example is shown in Figure 6.13.

As expected, for higher stirring rates, the average size of the capsules is lower (a finer emulsion is achieved during preparation, the polymerisation occurring at the surface of the lipophilic droplets). After the micro-encapsulation process, the microcapsules were vacuum filtered and then rinsed with deionised water. They were stored in deionised water for one day to separate the microcapsules from any dross (a mixture of broken

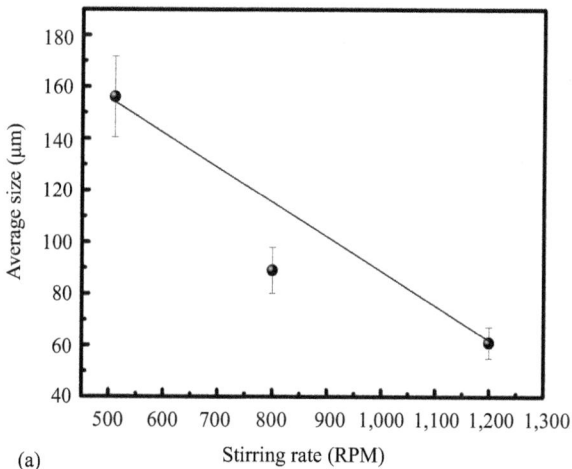

(a)

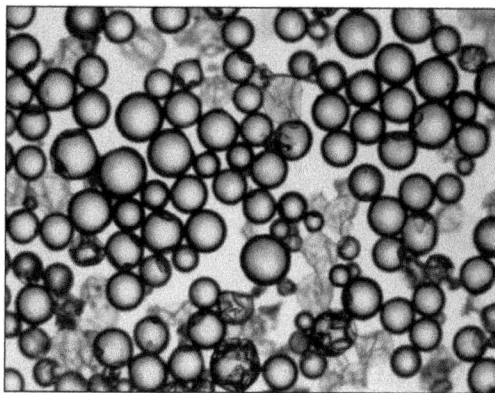

(b)

Figure 6.13   (a) The average size of the ENB microcapsules as a function of the stirring rate on rpm and (b) a typical optical photo of the ENB microcapsules prepared at 510 rpm

microcapsules and materials that did not form the capsules). The microcapsules were washed with water, and then rinsed with acetone under vacuum filtration, then kept under vacuum for about 5 minutes.

The microcapsules were then examined using optical microscopy. The microscopy showed that the dross disappeared completely after washing with water and acetone several times. A few shells of broken microcapsules were also observed, the microcapsules being rather fragile and rupturing easily in air. Weight loss of microcapsules was recorded using thermogravimetric analysis (TGA) when the temperature is kept at 30 °C. It showed that the weight of microcapsules drops very fast during the first 100 minutes, after that the loss tendency somehow becomes slow. After 2,500 minutes, the total weight loss is about 80%. The residue of the microcapsules is shown in Figure 6.14. Most of them are ruptured, and the ENB has completely evaporated.

The weight loss of the microcapsules as a function of the temperature (at a heating rate of 10 °C/min) is shown in Figure 6.15. It was found that ENB evaporated increasingly from the microcapsules. Besides, a sharp drop in the weight loss occurred at 230 °C, which indicated that most of capsules were completely ruptured at this temperature.

After 300 minutes, the total weight loss was about 80%. The remaining material is the urea-formaldehyde shell, which means that weight loss is due to the evaporation of the ENB. The following was concluded:

- ENB microcapsules are not stable in air even at RT and
- The urea-formaldehyde shell material decomposes before the temperature reaches 300 °C.

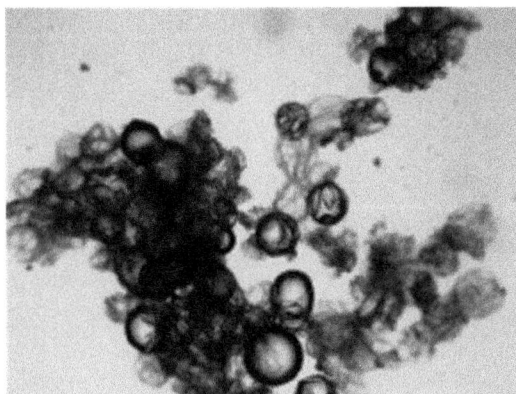

Figure 6.14   Optical image of the ENB monomer microencapsulated in PUF, in air after 2,500 minutes

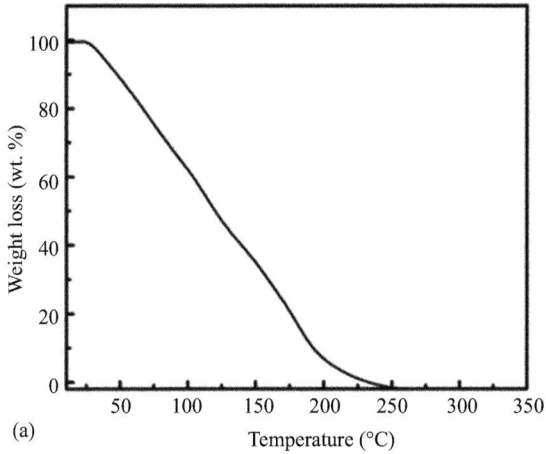

(a)

(b)

Figure 6.15 (a) ENB microcapsules weight loss with respect to the temperature (heating rate 10 °C/min) and (b) optical photo of the ENB microcapsules heated at 180 °C

6.3.2 Preparation of ENB microcapsules with polymelamine-urea-formaldehyde shells

Previous studies have pointed out a serious stability issue of microcapsules made with PUF shells. This is because the PUF shell is porous and rubbery, which causes formaldehyde emission. In addition, the shells walls are rather thin (160–220 nm thickness). In order to address this problem, poly-melamine-urea-formaldehyde (PMUF) was used as a substitute for PUF. The flowchart for the preparation of ENB/PMUF is shown in Figure 6.16.

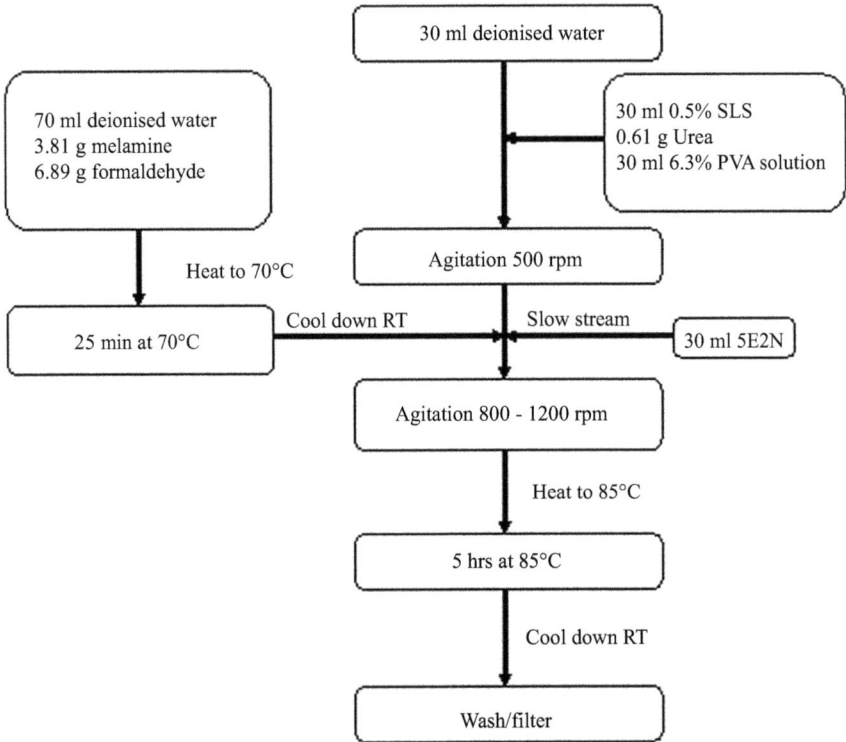

Figure 6.16   Flowchart for preparing ENB/PMUF microcapsules

Numerous runs were carried out to verify the properties of the sub-stitute. The average size of the microcapsules is controlled by the agitation speed of the mixer. It was observed that for relatively large microcapsule (greater than 100 μm) batches, the separated individual microcapsules are easily obtained by regular washing and filtering with water and acetone. With decreasing size, the tendency of agglomeration among the micro-capsules increases rapidly and it becomes extremely difficult to individually separate them out by standard washing techniques. Even some modified washing methods (such as washing in a sonicating bath, vacuum centrifu-ging and so on) did not contribute much to separating the individual microcapsules. Alternatively, modifying the amount and concentration of the emulsifying agent, polyvinyl alcohol (PVA) and sodium lauryl sulphate surfactant (SLS) in the microencapsulation process, contributed effectively in separating the individual microcapsules, especially when the average size is less than 20 μm.

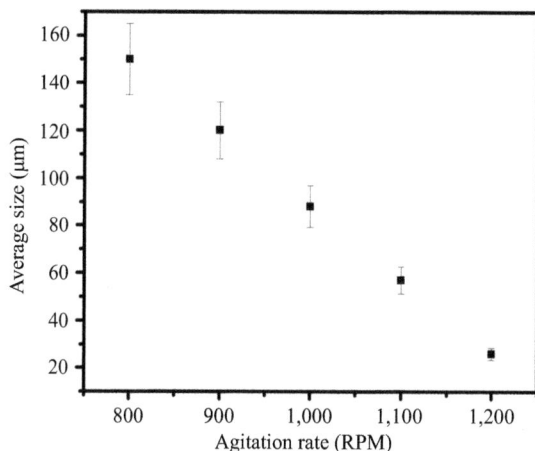

Figure 6.17   Effect of agitation speed on the size of the ENB/PMUF microcapsules

Numerous trials in synthesising the microcapsules were carried out by varying both the amount and the concentration of PVA and SLS, in order to establish an optimum condition for producing separated small microcapsules with an average size of less than 20 μm. The best quality capsules (individually separated) with an average size of less than 15 μm were found when 38 ml of 2 wt% SLS was used in combination with 22 ml of 6.3 wt% PVA, while keeping the other constituents the same. The effect of the agitation speed on the average size of the microcapsule is shown in Figure 6.17.

However, even with 1,200 rpm, the average size of the microcapsules was found to be in the range of 20 μm. With the aim of synthesising smaller microcapsules (in the nanscale range) that exhibit maximum toughening at lower concentrations, an *in situ* ultrasonication step during the synthesis process was introduced. Thus, ENB/PMUF microcapsules were prepared using an ultrasonic bath at different levels to create the emulsion during polymerisation. As for the stirring rate, the size of the microcapsules was inversely proportional to the applied ultrasound level. For example, for a power output of 9 W applied for 20 minutes (at 1,200 rpm), most of the capsules were between 0.5 and 2 μm in diameter, as shown in Figure 6.18. The size and wall thickness of these small microcapsules (less than 5 μm) were measured in dry state using SEM.

TGA was carried out to estimate the typical content of monomer inside the microcapsules, which was found to be around 80 wt%. This is shown in Figure 6.19.

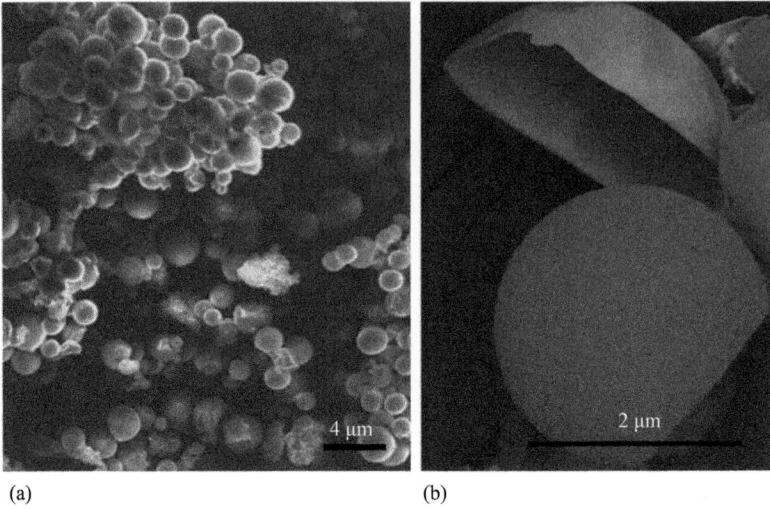

(a)   (b)

Figure 6.18   (a) SEM micrograph of small microcapsules and (b) SEM image showing the core/shell structure of the capsule

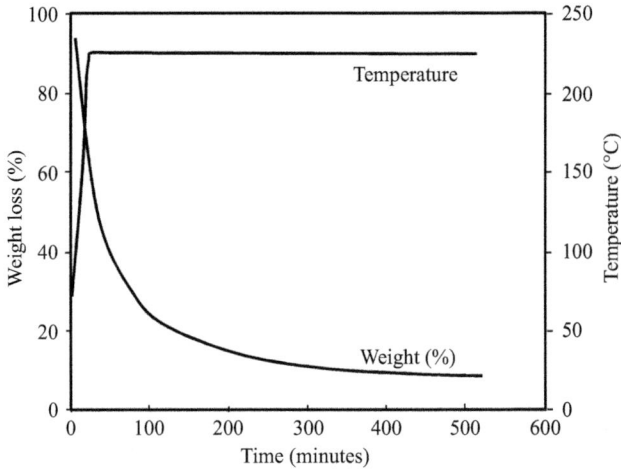

Figure 6.19   TGA results for estimating the monomer content inside the PMUF/ENB microcapsules

6.3.3   Comparison of the open-air stability of the polyurea-formaldehyde and polymelamine-urea-formaldehyde shells encapsulating ENB healing agent

The ENB/PUF microcapsules are not stable in air when their weight loss is about 55% within 2 hours at 30 °C (Figure 6.20). Changing from PUF to

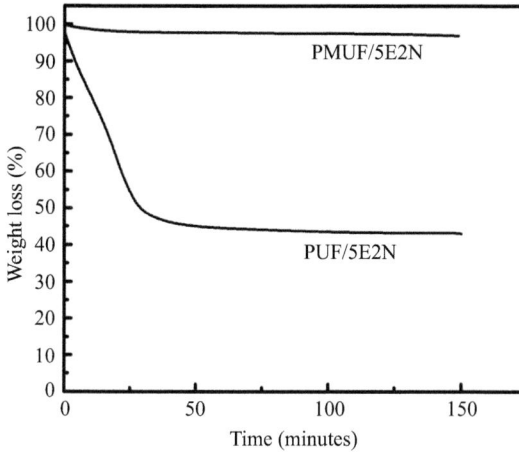

Figure 6.20 Weight loss of the ENB/PUF and ENB/PMUF microcapsules (at RT) as a function of time

PMUF for the shell's material greatly improves their stability. Indeed, the observed weight loss for the PMUF microcapsules is less than 3% even after 200 minutes at 30 °C.

6.4 Integration of the ENB monomer with single-walled carbon nanotubes into a microvascular network configuration

In this section, the fabrication of self-healing nanocomposite materials that consist of SWCNT reinforced ENB healing agent, reacted with RGC are described. The nanocomposite was produced by use of an ultrasonication process, followed by mixing on a three-roll mixing mill. The kinetics of the ENB-ROMP were studied as a function of the reaction temperature and the SWCNT loading. Besides, the microindentation analysis performed on the SWCNT/ENB nanocomposite materials after its ROMP showed a clear increase in both the hardness and the Young's modulus, that is, up to nine times higher than that of the virgin polymer, when SWCNT loading ranged from only 0.1 to 2 wt% [36]. This approach opens new prospects for using carbon nanotube (CNT) and healing agent nanocomposite materials for self-repair functionality, especially in the space environment.

6.4.1 Experimental details

The SWCNT materials were synthesised by using plasma torch technology (the detailed process can be found in [37,38]). In this approach, a carbon-containing ethylene (C_2H_4) substance combined with a gaseous catalyst based on ferrocene [$Fe(C_5H_5)_2$] vapour are injected into an inert gas plasma jet. The inert gas is a mixture of 50% argon with 50% helium. The hot temperature of the plasma ($\sim$4727 °C) is capable of dissociating the ethylene and ferrocene molecules to produce iron and carbon vapour. These atomic and molecular species are then rapidly cooled at a rate of 10^4 °C/s in a warmer environment of about 1000 °C. The process produces SWCNT materials, and the growth takes place in the gas phase. The as-grown, soot-like SWCNT were purified by an acidic treatment of refluxing in a 3 M HNO_3 (Sigma–Aldrich) solution for 5 hours at 130 °C, and subsequently filtered (Filter type GV, Millipore Corp).

The plasma grown CNTs were characterised by SEM using a Jeol JSM-6300F microscope. Bright-field transmission electron microscopy (TEM) images were obtained using a Jeol JEM-2100F FEG-TEM (200 kV) microscope. The Raman measurements were performed by the 514.5 nm (2.41 eV) laser radiation of an Ar^+ laser focused onto the sample with a spot of 1 μm (microRaman spectroscopy, Renishaw Imaging Microscope WiR-E^{TM}). The Raman spectra were taken with a back scattering geometry at RT in the 100–2,000 cm^{-1} spectral region.

All the chemicals [first generation RGC, ENB monomer, Epon^{TM} 828 resin epoxy (Hexion Chemicals) and so on] were purchased from Sigma–Aldrich and used as received. The weighed amount of purified SWCNT was first dispersed in a solution of ENB. The resulting nanocomposites containing a different mass fraction of SWCNT were passed several times through a three-roll mixing mill (Exakt 80 E, Exakt Technologies Inc.) where the gap between the rolls and the speed of the apron roll was adjusted according to the method described by Thostenson and Chou [39]. The total procedure consisted of five passes through a gap of 25 μm and with a speed of 200 rpm, five passes through a gap of 15 μm with a speed of 200 rpm and finally nine passes through a gap of 5 μm with a speed of 250 rpm. The nanocomposite solution was subjected to an ultrasonication step (ultrasonic cleaner 8891, Cole-Parmer) for 30 minutes. In parallel, an RGC powder (Sigma–Aldrich) was mixed with an organic solution of dichloromethane (Sigma–Aldrich) and subjected to the same mechanical blending process as the SWCNT. After solvent evaporation, the fine RGC powder was slowly added to the nanocomposite solution containing SWCNT with gentle stirring (Barnstead hotplate stirrer, model SP131825, Barnstead International), and the nanocomposite samples obtained were immediately placed in a Tenney Junior Environment Chamber^{TM} under air, where the time of polymerisation was evaluated starting from this final step. Doing so, the ROMP reactions were then performed at different temperatures ranging from −15 to 45 °C.

The microscopic scale dispersion was characterised by observing a 1 mm thick film under a transmission light optical microscope using the 40× objective (BX61, Olympus) and image analysis software (Image-Pro Plus, Media Cybernetics Inc.).

After their full polymerisation, the mechanical properties of the ENB/ SWCNT nanocomposites obtained were characterised by depth sensing indentation using a commercially available CSM Micro Indentation Tester (CSM Instruments), equipped with a Vickers diamond tip. The applied load was 3N. Hardness (*H*) and Young's modulus (*E*), for each sample, were obtained from a minimum of 10 indentations tests.

6.4.2   Results and discussion

Figure 6.21(a) shows a representative TEM micrograph of the as-grown SWCNT deposit, where bundles of a few single-walled nanotubes (the diameter of the individual tubes is about 1.2 nm) are clearly seen. In conjunction with the SWCNT, other carbon nanostructures (e.g., nanocages, nanonions and nanohorns) can be also produced by the plasma torch process. Metal catalysts NP were also expected to be present in the as-grown deposits (as shown by the black arrows in Figure 6.21(a)). The nanotubes purification eliminates a large part of the catalyst residues and other carbon nanostructures having a high density of structural defects. The purified SWCNT consist of bundles having diameters in 2–10 nm range and lengths in the order of few μm which have aspect ratios of at least three orders of magnitude. A typical Raman spectrum (Figure 6.21(b)) of the purified nanotube material shows clear scattering peaks appearing in the low (100–300 cm^{-1}) and high ($\sim 1,600$ cm^{-1}) frequency regions, corresponding to the radial breathing mode (RBM) and the tangential vibrating mode (G), respectively, which are the fingerprints of the presence of SWCNT. The RBM peak centred at 185 cm^{-1} is attributed to the strong presence of SWCNT having a mean diameter of 1.2 nm according to the formula of Bandow *et al.* [40]. The disordered peak (D-peak) centred around $1,350$ cm^{-1} is due to the presence of amorphous and/or disordered carbon structures. Nevertheless, the very low D-to-G peak intensity ratio (~ 0.05) is worth noting, as it indicates the overall high quality of the SWCNT mats. The TEM observations are consistent with the Raman results, verifying the single-walled structure of the CNT grown as well as their narrow diameter.

To demonstrate the kinetics of the polymerisation reaction as a function of the temperature and the SWCNT concentrations, as was done in the study described in Section 6.1, ROMP reactions tests were performed at temperatures ranging from -15 to 45 °C and CNT loads ranging from

(a)

(b)

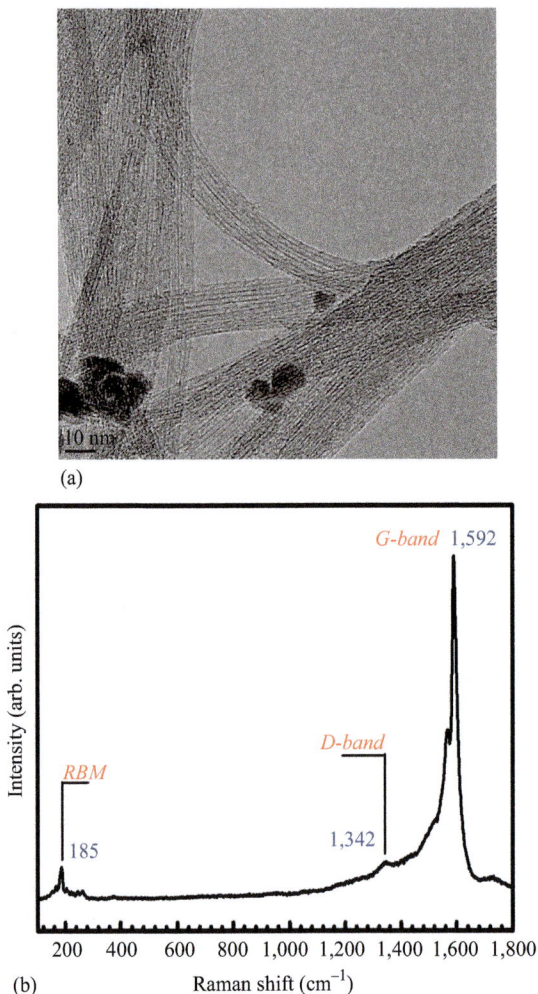

Figure 6.21   *Morphology of the grown SWCNT (a) representative TEM*
images of the purified SWCNT materials and (b) typical
Raman spectrum of the SWCNT mats, where the various
RBM, D and G bands are clearly identified
[Reproduced, with permission, from [36], ©2012 IOP Publishing Ltd]

0 to 5 wt%. Figure 6.22 illustrates the time needed for the ENB polymerisation (ROMP) as a function of the reaction temperature. This time is
found to be 146 minutes at −15 °C, which decreases to 4 minutes only
when the reaction temperature increases to 20 °C. The polymerisation then
occurs in a very short time (as low as 0.2 minutes) when the temperature is
increased to 45 °C. It is worth noting here that no significant change in
the kinetics of the ROMP reaction was observed with respect to the
CNT loads. So, at RT, the ROMP reaction of the ENB monomer triggered

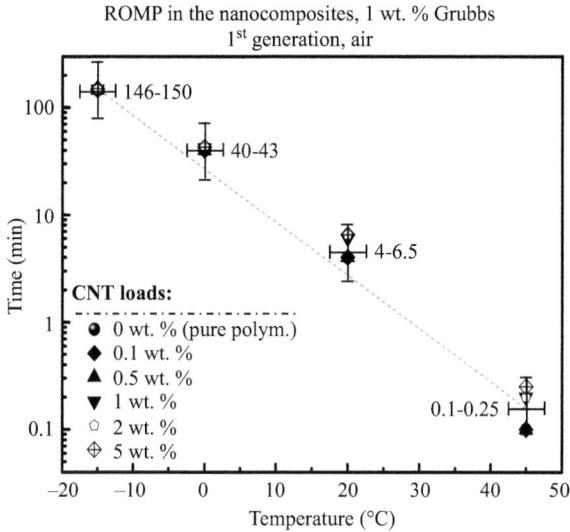

Figure 6.22 Time of the ENB polymerisation (ROMP) as a function of reaction temperature and CNT loads (1st generation Grubbs' catalyst, 1 wt%, in air)
[Reproduced, with permission, from [36], © 2012 IOP Publishing Ltd]

by the Grubbs' catalyst is found to happen in a very short time (less than 5 minutes). The ENB monomer definitely provided the best polymerisation time compared to alternative monomers [26].

Micro/nanoindentation is now commonly used to investigate the mechanical properties of materials at the micro/nanoscale. This is a straightforward technique, which can be used to measure the mechanical properties of the bulk of the polymer and thin films. The microhardness test is based on the standards for instrumented indentation, ASTM E2546 [41] and ISO 14577-1 [42]. It uses an established method where an indenter tip with a known geometry is driven into a specific site of the material to be tested, by applying an increasing normal load. When reaching a preset maximum value, the normal load is reduced until partial or complete relaxation occurs. This procedure is performed repetitively; at each stage of the experiment the position of the indenter relative to the sample surface is precisely monitored with an optical noncontact depth sensor.

For each indentation, the applied load value was plotted with respect to the corresponding position of the indenter. The resulting load/displacement curves provide data specific to the mechanical nature of the material under examination, and from these load-displacement data, many mechanical properties such as hardness and elastic properties (*E*) can be determined [43]. In short, the mechanical properties are determined based on the mechanical response of the specimen to the applied load as well as the indenter geometry [43–45].

A hardness tester with a Vickers diamond shape indenter was used. Established models were applied to calculate quantitative hardness and modulus values for such data. In fact, the system was equipped with an automatic computation method of Oliver and Pharr [46] with its specific software directly giving the mechanical properties.

The hardness and the Young's modulus are the two main mechanical properties frequently determined using indentation techniques. As the indenter is pressed into the sample, both elastic and plastic deformation occurs and a hardness (H) impression conforming to the indenter geometry is formed. Only the elastic portion of the displacement is recovered during indenter withdrawal, which enables the calculation of elastic modulus. Micro/nanohardness is defined as the indentation load divided by the estimated contact area of the indentation. The static mechanical characterisation of CNT/ENB nanocomposites was performed with this technique [45]. The mechanical properties of a CNT/ENB nanocomposite having CNT loads ranging from 0 to 5 wt% were characterised. A typical example of the load-displacement curves obtained for a 3 N load is shown in Figure 6.23(a) while the postindentation residual impression optical images of pure ENB (i.e., ENB monomer reacted with RGC and 0 wt% of CNT) and ENB loaded with 2 wt% of CNT are shown in Figure 6.23(b) and Figure 6.23(c), respectively.

All the mechanical properties obtained are summarised in Table 6.7. When higher loads were applied, both H and E increased with the CNT load. Compared to pure ENB polymer (i.e., for 0 wt% CNT), analysis of microindentation results show that CNT/ENB exhibits a higher elastic strain to failure and resistance to plastic deformation, expressed by the H/E and the H^3/E^2 ratios, respectively.

Table 6.7 summarises the hardness (H), Young's modulus (E), elastic strain (H/E) and the resistance to plastic deformation (H^2/E^2) of the neat ENB polymer and its nanocomposite samples as a function of the nanotube contents. A clear improvement in all mechanical properties, compared to those of neat ENB samples, was obtained for CNT loads as low as 0.1 wt% and continued to increase with increase in the nanotube content (e.g., an enhancement as high as 900% is obtained for hardness for a CNT load of 2 wt% only), indicating the unique reinforcing effects of nanotubes [45].

The incorporation of CNT fillers may act as a crosslinking network inside the ENB polymer and consequently leads to an increase of its overall mechanical properties. However, it is worth noting that at a CNT loading of 5 wt%, samples do not show any further improvement when compared to those observed for the 0.1–2 wt% interval. These results can be explained by the fact that the high surface area of the CNT increases with their concentration in the nanocomposite. At 5 wt%, there it is highly likely that there is less available polymer to be intercalated into the CNT bundles.

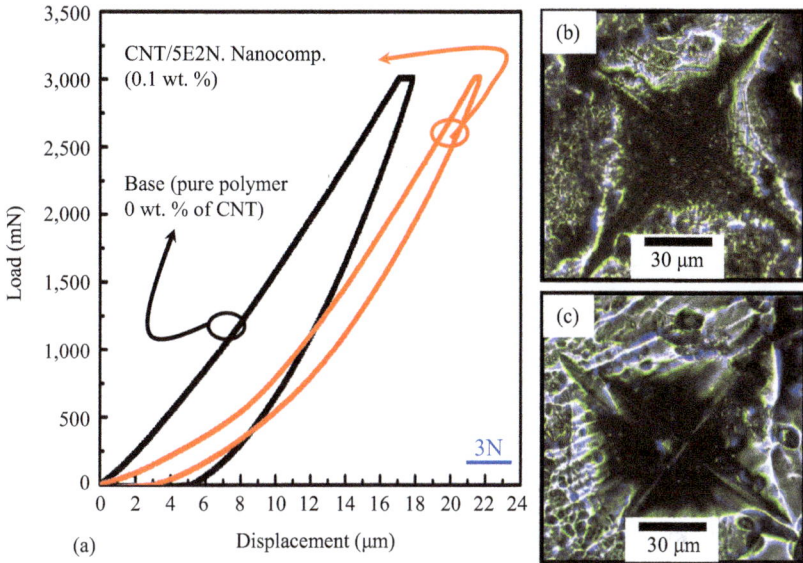

Figure 6.23 *(a) Representative load-displacement curves of indentations made at a peak indentation load of 3 N on ENB and its SWCNT reinforced samples. Postindentation residual impressions: (b) pure ENB reacted with RGC (i.e., 0 wt% CNT) and (c) ENB/CNT (CNT load of 2 wt%) taken at 3 N*

Table 6.7 Summary all the mechanical properties obtained from Vickers microindentation at a fixed applied load of 3 N

CNT loads	Hardness (H) (GPa)	Young's modulus, (E) (GPa)	Elastic strain to failure (H/E)	Resistance to plastic deformation (H^2/E^2) (GPa)
Virgin sample (0 wt%)	0.4	3.6	0.11	0.0123
Load 1: 0.1 wt%	1.3	5.9	0.22	0.0485
Load 2: 0.5 wt%	1.9	8	0.23	0.0564
Load 3: 1 wt%	2.7	11	0.24	0.0602
Load 4: 2 wt%	3.6	14	0.25	0.0661
Load 5: 5 wt%	0.8	4.6	0.17	0.0302

Consequently, the interactions between the nanotubes are much higher at this level of loading, giving rise to the formation of aggregates. The aggregation of nanotube bundles decreases the effective content of SWCNT in the polymer matrix. At this load, these aggregates may cause SCWNT pull-out, break-off, break-out or snagging in the samples that are directly responsible for the mechanical properties obtained – which are still less than their theoretically predicted potential [47]. Besides, the quantity of aggregates increases with the nanotube concentration, preventing further interaction of the polymer with the SWCNT [44], and probably, only the outside nanotubes of a bundle can be bonded to the polymer matrix. The inside nanotubes are weakly interacting by the van der Waals attraction. The nanotubes within a bundle can easily slide past each other and the shear modulus of the CNT bundles is relatively low [48]. This is one of the reasons why the mechanical properties of the higher SWCNT concentrations are significantly decreased [47]. For low nanotube concentrations (i.e., less than 2 wt%), an intercalation of the polymer inside the SWCNT bundles could be enabled, thus, helping the nanotube dispersion. Thus, interactions between nanotubes are low and the bundles can be desegregated. Thus, the existence of CNT aggregates at higher concentrations is believed to be responsible for the relatively low hardness and Young's modulus, as compared to that expected from theoretically predicted values [47]. It is worth noting here that the theoretical predictions of strength and elastic modulus are almost all for single nanotubes that are well dispersed in a matrix, whilst calculations based on experiments have not yet well addressed the possibility of relative slippage of the individual tubes within a bundle as well as the aggregation of CNT bundles. Because of this, these hypotheses have to be treated with great care, and more quantitative characterisations (such as processing the cryo-microtome technique as a function of the CNT loads) are needed to support these claims. Intercalation of polymers into the bundles is one of the key reinforcing mechanisms at work in the SWCNT polymer nanocomposites, and better dispersion techniques are inevitably needed to overcome such limitations as those mentioned previously. Finally, the fact that the mechanical properties start to change for a CNT load as low as 0.1 wt% suggests that it is highly likely that both mechanical and/or the electrical percolation thresholds could be around this value.

6.4.3   Elaboration of the three-dimensional microvascular network and self-healing testing

Three-dimensional microscaffolds were fabricated using a computer-controlled robot (I&J 2200-4, I & J Fisnar) that moves a dispensing apparatus (HP7X^{TM}, EFD) along the x, y and z axes [49]. The fabrication of the microscaffold began with the deposition of the ink-based filaments on an epoxy substrate, leading to a two-dimensional pattern. The following layers were deposited by successively incrementing the z-position of the dispensing nozzle by the diameter of the filaments. The 3D microscaffold consisted of 11 layers of fugitive ink filaments, deposited alternatively along and perpendicular to the scaffold longitudinal, x-axis. The filament diameter was 150 μm for a deposition speed of 4.7 mm/s at an extrusion pressure of 1.9 MPa. The overall dimensions of the 3D ink structure were 62 mm in length, 8 mm in width and 1.7 mm in thickness with 0.25 mm spacing between filaments. The empty space between the scaffold filaments was filled with the same epoxy resin used for the substrate fabrication (i.e., Epon™ 828/Epikure™ 3274, Miller-Stephenson Chemical Co.), mixed with the fine powder of the Grubbs' catalyst particles. After the curing of the epoxy, the fugitive ink was removed from the structure by liquefaction at 100 °C and application of a vacuum, which yielded an interconnected 3D microfluidic network. Figure 6.24 schematically illustrates this fabrication process.

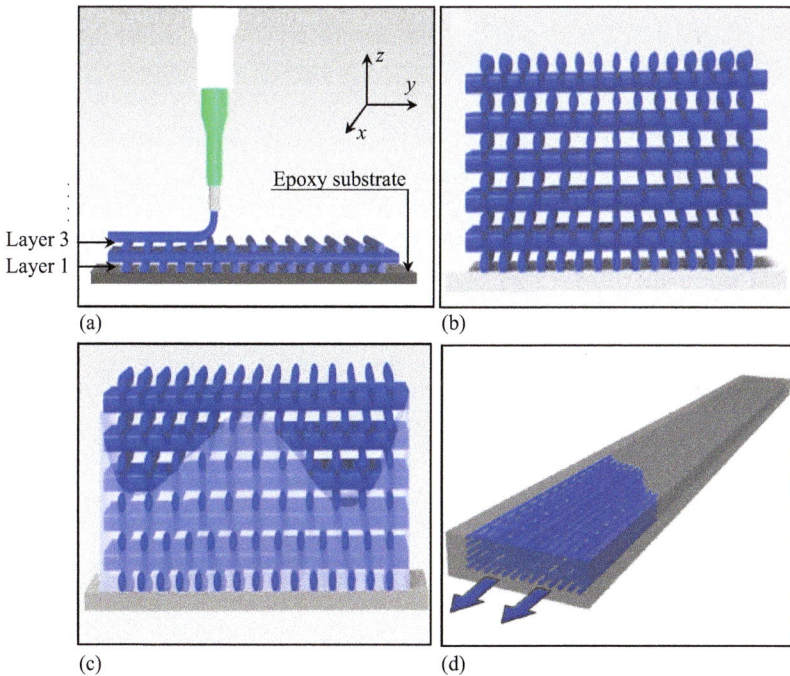

Figure 6.24   *Schematic representation of the manufacturing process of a 3D-reinforced nanocomposite beam through microinjection of 3D microfluidic network: (a) and (b) deposition of fugitive ink scaffold on an epoxy substrate; (c) encapsulation of the 3D ink-based scaffold using epoxy resin containing Grubbs' catalyst followed by resin solidification and (d) ink removal at 100 °C under vacuum*
    [Reproduced, with permission, from [36], © 2012 IOP Publishing Ltd]

The 3D-reinforced beams were produced by microinjecting the empty microfluidic networks with the liquid nanocomposites (ENB/CNT) having a CNT load of 0.5 wt%. The materials injected behaved like microscale fibres inside a matrix of epoxy/RGC-NP. Figure 6.25 schematically illustrates the microinjection step for a nanocomposite-injected beam. The nanocomposites were injected into the empty channels using a fluid dispenser (EFD 800, EFD) *via* a plastic tube connected to the end of the beams and the fluid dispenser at both ends. The injection pressure was adjusted at 400 kPa, which led to an injection speed of ~1 mm/s. The final specimen was then sealed to protect the liquid nanocomposite inside the sample.

Figure 6.26(a) shows an optical, top-view image of a microvascular nanocomposite infiltrated network just after the impact damage (hole), caused by a low velocity dropping mass (impact energy ~23 J). In Figure 6.26(b), we can see the microvascular sample heated at 60 °C for 15 minutes, where the impact damage seems to be totally filled by the liquid ENB/SWCNT nanocomposite healing agent, and is completely solidified after 30 minutes (Figure 6.25(c)). The liquid ENB/SWCNT nanocomposite was released after the impact event and met the RGC-NP already embedded

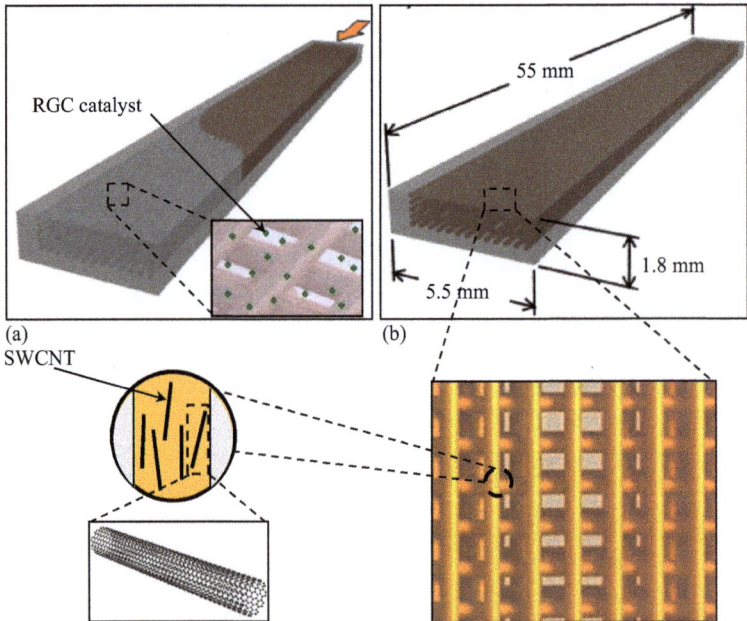

Figure 6.25   (a) Microinjection of the empty network by the prepared liquid ENB/SWCNT nanocomposite and (b) cut beam to final dimensions
[Reproduced, with permission, from [36], © 2012 IOP Publishing Ltd]

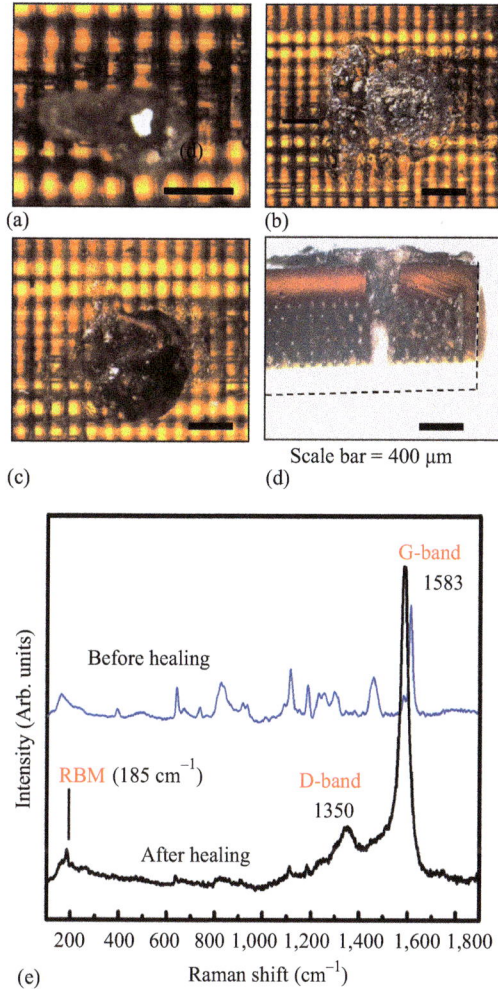

Figure 6.26 (a) *Optical top-view image of a microvascular nanocomposite infiltrated network just after the impact event; (b) sample heated at 60 °C for 15 minutes; (c) sample after 30 minutes at 60 °C where the liquid healing-agent was completely solidified; (d) cross-sectional view of the healed damage in (c) showing the clear polymerisation inside the hole and (e) Raman spectrum performed directly on the damage zone before impact event and after hole repairing, showing clear presence of the SWCNT materials (see RBM, G and D bands) in the self-healed damage*
[Reproduced, with permission, from [36], © 2012 IOP Publishing Ltd]

into the epoxy scaffold where the ROMP reaction has taken place. It can clearly be seen in the cross-sectional view (Figure 6.26(d)), the polymerisation of the ENB/SWCNT nanocomposite inside the damage-zone. On the other hand, the Raman spectrum performed directly on the damage-zone, before the impact event (i.e., on the undamaged epoxy scaffold) and after the complete healing process (where the damage hole was completely filled and completely solidified) is shown in Figure 6.26(e). The presence of the SWCNT materials in the repaired damage is clearly confirmed by the RBM, D and G Raman peaks [50,51].

After this proof of concept step, it can be seen that the self-healing system in this study has the potential to be used in practical applications. Self-healing has most benefit for structures where good durability and long service life are important, or for structures that are difficult to reach for inspection and maintenance. More work will be done to characterise mechanically the healing efficiency of our microvascular devices as a function of the nanotube loads, the damage morphology (i.e., the geometry and the shape of the impact zone), together with the dynamics of the healing reaction itself [52].

The results obtained definitely open up new prospects for using CNT and healing agent nanocomposite materials with higher mechanical characteristics for self-repair functionality, especially in the space environment.

References

[1] B. Aïssa, R. Nechache, E. Haddad, W. Jamroz, P.G. Merle and F. Rosei, *Applied Surface Science*, 2012, **258**, 24, 9800.
[2] R. Eason, Ed., *Pulsed Laser Deposition of Thin Films: Applications-Led Growth of Functional Materials*, Wiley, Hoboken, NJ, 2006.
[3] H.M. Smith and A.F. Turner, *Applied Optics*, 1965, **4**, 1, 147.
[4] D. Dijkkamp, T. Venkatesan, X.D. Wu, *et al.*, *Applied Physics Letters*, 1987, **51**, 8, 619.
[5] D.B. Chrisey and G.K. Hubler, Eds., *Pulsed Laser Deposition of Thin Films*, John Wiley, New York, NY, 1994.
[6] C.R. Guarnieri, R.A. Roy, K.L. Saenger, S.A. Shivashankar, D.S. Yee and J.J. Cuomo, *Applied Physics Letters*, 1988, **53**, 6, 532.
[7] C.M. Foster, K.F. Voss, T.W. Hagler, *et al.*, *Solid State Communications*, 1990, **76**, 5, 651.
[8] S.R. Shinde, S.B. Ogale, R.L. Greene, *et al.*, *Applied Physics Letters*, 2001, **79**, 2, 227.
[9] E. Fogarassy, C. Fuchs, A. Slaoui and J.P. Stoquert, *Applied Physics Letters*, 1990, **57**, 7, 664.

[10] M. Balooch, R.J. Tench, W.J. Siekhaus, M.J. Allen, A.L. Connor and D.R. Olander, *Applied Physics Letters*, 1990, **57**, 15, 1540.

[11] N. Biunno, J. Narayan, S.K. Hofmeister, A.R. Srivatsa and R.K. Singh, *Applied Physics Letters*, 1989, **54**, 16, 1519.

[12] H. Kidoh, T. Ogawa, A. Morimoto and T. Shimizu, *Applied Physics Letters*, 1991, **58**, 25, 2910.

[13] J.A. Martin, L. Vazquez, P. Bernard, F. Comin and S. Ferrer, *Applied Physics Letters*, 1990, **57**, 17, 1742.

[14] R.E. Smalley and R.F. Curl, *Scientific American*, 1991, **265**, 4, 32.

[15] S.G. Hansen and T.E. Robitaille, *Applied Physics Letters*, 1988, **52**, 1, 81.

[16] H-U. Krebs and O. Bremert, *Applied Physics Letters*, 1993, **62**, 19, 2341.

[17] A.J.M. Geurtsen, J.C.S Kools, L. de Wit and J.C. Lodder, *Applied Surface Science*, 1996, **96–98**, 887.

[18] R. Nechache, C. Harnagea, A. Pignolet, *et al.*, *Applied Physics Letters*, 2006, **89**, 10, 102902.

[19] S. Fähler and H-U. Krebs, *Applied Surface Science*, 1996, **96–98**, 61.

[20] R. Kelly and J.E. Rothenberg, *Nuclear Instruments and Methods in Physics Research: B*, 1985, **7–8**, 2, 755.

[21] C.R. Phipps, Jr., T.P. Turner, R.F. Harrison, *et al.*, *Journal of Applied Physics*, 1988, **64**, 3, 1083.

[22] C.W. Bielawski and R.H. Grubbs, *Progress in Polymer Science*, 2007, **32**, 1, 1.

[23] X. Liu, X. Sheng, J.K. Lee, M.R. Kessler and J.S. Kim, *Composites Science and Technology*, 2009, **69**, 13, 2102.

[24] S.R. White, N.R. Sottos, P.H. Geubelle, *et al.*, *Nature*, 2001, **409**, 6822, 794.

[25] A.S. Jones, J.D. Rule, J.S. Moore, N.R. Sottos and S.R. White, *Journal of the Royal Society – Interface*, 2007, **4**, 13, 395.

[26] R.P. Wool, *Soft Matter*, 2008, **4**, 3, 400.

[27] R.S. Trask, G.J. Williams and I.P. Bond, *Journal of the Royal Society – Interface*, 2007, **4**, 13, 363.

[28] X. Liu, J.K. Lee, S.H. Yoon and Michael R. Kessler, *Journal of Applied Polymer Science*, 2006, **101**, 3, 1266.

[29] M.R. Kessler and S.R. White, *Composites Part A: Applied Science and Manufacturing*, 2001, **32**, 5, 683.

[30] E.N. Brown, N.R. Sottos and S.R. White, *Experimental Mechanics*, 2002, **42**, 4, 372.

[31] E.N. Brown, M.R. Kessler, N.R. Sottos and S.R. White, *Journal of Microencapsulation*, 2003, **20**, 6, 719.

[32] M.R. Kessler, N.R. Sottos and S.R. White, *Composites Part A: Applied Science and Manufacturing*, 2003, **34**, 8, 743.

[33] L. Yuan, A. Gu and G. Liang, *Materials Chemistry and Physics*, 2008, **110**, 2–3, 417–425.

[34] A.C. Jackson, B.J. Blaiszik, D. McIlroy, N.R. Sottos and P.V. Braun, *Polymer Preprints*, 2008, **49**, 1, 967.

[35] R.S. Trask and I.P. Bond, *Smart Materials and Structures*, 2006, **15**, 3, 704.

[36] B. Aïssa, E. Haddad, W. Jamroz, *et al.*, *Smart Materials and Structures*, 2012, **21**, 10, 105028.

[37] O. Smiljanic, B.L. Stansfield, and J-P. Dodelet, A. Serventi and S. Désilets, *Chemical Physics Letters*, 2002, **356**, 3–4, 189.

[38] O. Smiljanic, F. Larouche, X.L. Sun, J-P. Dodelet and B.L. Stansfield, *Journal of Nanoscience and Nanotechnology*, 2004, **4**, 8, 1005.

[39] E.T. Thostenson and T-W. Chou, *Carbon*, 2006, **44**, 14, 3022.

[40] S. Bandow, S. Asaka, Y. Saito, *et al.*, *Physical Review Letters*, 1998, **80**, 17, 3779.

[41] ASTM E2546, *Practice for Instrumented Indentation Testing*, 2007.

[42] ISO 14577-1, *Metallic Materials – Instrumented Indentation Test for Hardness and Materials Parameters – Part 1: Test Method*, 2003.

[43] X.D. Li and B. Bhushan, *Materials Characterization*, 2002, **48**, 1, 11.

[44] L. Valentini, J. Biagiotti, J.M. Kenny and M.A.L. Manchado, *Journal of Applied Polymer Science*, 2003, **89**, 10, 2657.

[45] X. Li, H. Gao, W.A. Scrivens, *et al.*, *Nanotechnology*, 2004, **15**, 11, 1416.

[46] W.C. Oliver and G.M. Pharr, *Journal of Materials Research*, 1992, **7**, 6, 1564.

[47] A.A. Mamedov, N.A. Kotov, M. Prato, D.M. Guldi, J.P. Wicksted and A. Hirsch, *Nature Materials*, 2002, **1**, 3, 190.

[48] P.C.P. Watts, W.K. Hsu, G.Z. Chen, D.J. Fray, H.W. Kroto and D.R.M. Walton, *Journal of Materials Chemistry*, 2001, **11**, 10, 2482.

[49] D. Therriault, S.R. White and J.A. Lewis, *Nature Materials*, 2003, **2**, 4, 265.

[50] M.S. Dresselhaus, A. Jorio, A.G. Souza Filho, G. Dresselhaus and R. Saito, *Physica B: Condensed Matter*, 2002, **323**, 1–4, 15–20.

[51] S. Suzuki and H. Hibino, *Carbon*, 2011, **49**, 7, 2264–2272.

[52] S. Mostovoy, P.B. Crosley and E.J. Ripling, *Journal of Materials*, 1967, **2**, 3, 661.

Chapter 7

Self-healing in space environment

The space environment is quite hostile to structural materials. Far from being just an inert vacuum, the low Earth orbit (LEO) situated at an altitude of 200–700 km contains reactive atomic oxygen (AO), an increasing quantity of man-made debris, natural micrometeoroids, ultraviolet (UV) radiation, electromagnetic radiation, particulate radiation (electrons, protons and heavy ions) and large extremes of temperature. This is why the lifetime of spacecraft is determined by the environmentally induced degradation of the structural materials. The present generations of satellites and spacecrafts have a life expectancy, which is limited to five to seven years.

Therefore, it is highly desirable to have self-healing systems, which can mitigate damage that could lead to catastrophic failures as follows:

- Failure of structural parts (e.g., the Challenger and Columbia spacecraft accidents),
- Failure of electrical wire insulation materials used in spacecrafts (NASA flight STS-93) and
- Failure of polymer membranes used in the space instruments.

For these reasons various self-healing approaches for space applications have been investigated.

The self-healing systems are applied mainly to polymer matrix-based materials. They usually include a variety of epoxies, polyimides, polysulfones and phenolics. In particular, cyanate-ester resins have been considered due to their lower hygrostrain/outgassing compared to the first generation of epoxy matrices. Carbon, glass and aramid fibres have also been used as reinforcing filaments within composite space structures. Aramid fibres are often employed as a 'buffer' within a shielding system against damage induced by micrometeoroid impacts. Examples of some composite materials for space are identified in Table 7.1.

Numerous experimental programmes have systematically been undertaken on Earth and within a space environment on dedicated research platforms (e.g., long duration exposure facility (LDEF)) and test facilities within manned flights (e.g., The National Aeronautics and Space Administration (NASA) Space Shuttle). These test programmes have been supplemented and

Table 7.1   Typical composite materials used in space structures

Material	Trade name and chemical
Carbon fibres	T300 fibres
	GY-70 (fibres, Celanese)
	P75
	HMF 176
	T50
	AS-4 fibre
Ceramic composites	SiC, ZrB_2 and Y_2O_3-based powders and composites
Resin type	
Epoxy	934 resin
	5028 resin
	X30
	CE 339
	HexPly® F263 (Hexcel)
	X904B
	3501-6 (Hexcel)
	ERL-1962

combined with additional information retrieved from composite structures returned to Earth (e.g., solar cells from the Russian Space Station, Mir). This wealth of information has considerably helped the designers to modify their approaches and existing terrestrial design concepts so they can meet the complex demands of using polymer-based composite materials in the space environment.

This knowledge has also helped to generate stringent verification test programme, which determine whether a composite material is deemed to be acceptable for the space environment or not. For example, for composite structures to be approved for the European Space Agency (ESA), the materials must pass the technical requirements outlined in the ECSS-Q-70-04 protocol [1] for thermal cycling and the ECSS-Q-70-02 protocol [2] for thermal vacuum testing.

A summary of the space environment in LEO is given in Table 7.2, whilst Table 7.3 defines the mission environment for circular low Earth orbits (CLEO) and highly elliptical orbits (HEO) [3].

The exact influence of each individual constituent of the space environment, as well as their effects, should be well understood before the selection of the composite and/or the self-healing resin system is undertaken. Certain aspects of the space environment will affect the composite structure in different ways.

Table 7.2 LEO space environment characteristics

Environment	LEO conditions
High vacuum	1.3×10^{-9} to 1.3×10^{-10} (Torr)
UV radiation	100–400 nm, 4,500–14,500 equivalent sun hours
Atomic oxygen	10^{-3} to 9.02×10^{21} atoms/cm^2 (motion-facing surface) after 5.8 years exposure on a long duration exposure facility (LDEF)
Meteoroid and debris impacts	>36,000 particles from ~0.1 to ~2.5 mm – high effect on ram-facing surfaces
Thermal cycling	LEO: -47 °C to 85 °C, ± 11 °C. Worst case: -160 to 160 °C
UV: Ultraviolet	

Table 7.3 Predicted mission environmental parameters

Space parameters	CLEO	HEO
Estimated life time electron radiation dose (Mrad)	10	1,000
Thermal cycle, °C	-100 ± 20 to $+100 \pm 20$	-150 ± 20 to $+150 \pm 20$
Lifetime, years	>10	>10
Orbit, nautical miles (NM)	<450	From 450 to 22,500
	28.5°	28.5°

7.1 Challenges of the self-healing reaction in the space environment

In general, the healing methods can be divided into four major categories:

- Capsules or particles randomly spread within the structure;
- An organised net based on hollow fibres or a microvascular system;
- An organised net based on wires (shape memory alloys, fibres and conductive metallic wires) and
- An organised net with an external triggering system (active system).

An example of self-healing with capsules is the repair process implemented within composite materials. In this system, a monomer is encapsulated and then dispersed together with a catalyst. Once the microcapsules break (rupture), the monomer flows within the crack and is polymerised by the dispersed catalyst. In this way the microfissure is repaired. Another approach is to add a triggering mechanism to a passive system. In such a case, the repair process is activated externally. For example, sunlight may serve as the triggering mechanism. At the same time, the sunlight may be used to improve the efficiency of the curing process.

Healing materials increase the safety, reliability and lifetime of airframe, launcher and space structures by reducing the propagation of fatigue damage and mitigating the growth of small cracks in the structural materials. However, the added healing agents affect the intrinsic host material properties, requiring a complete verification of the material's strength, manufacturing process and its lifetime.

Figure 7.1 summarises the taxonomy of the concepts, approaches and methods used to validate the self-healing technology.

The use of functional components stored inside composite materials to restore physical properties after damage has long been advocated by several authors. The first overview of self-healing materials was published

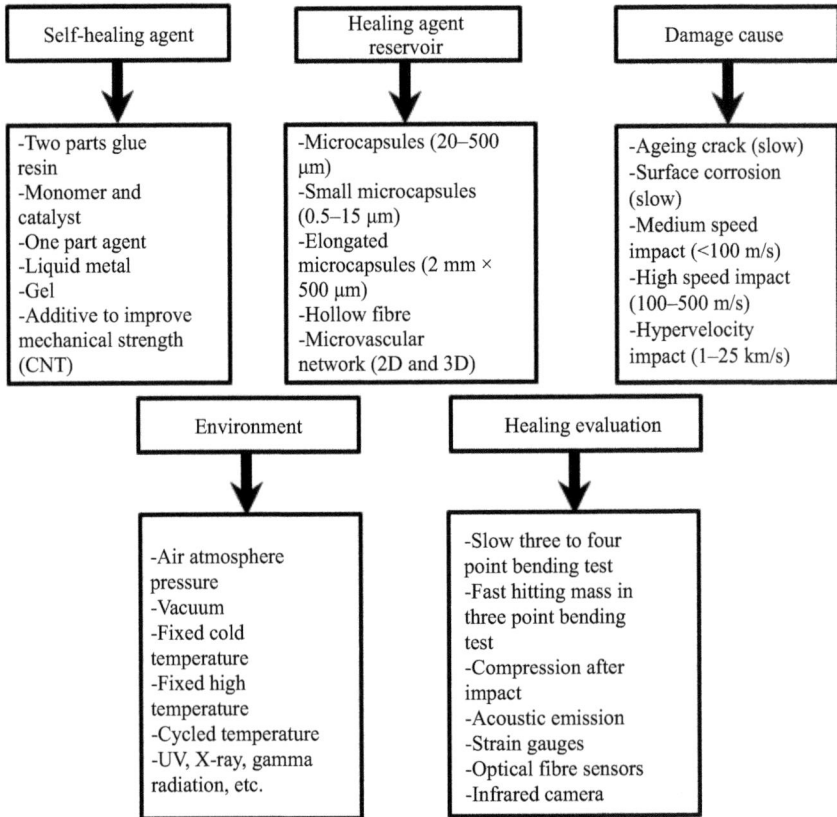

Self-healing agent	Healing agent reservoir	Damage cause
-Two parts glue resin -Monomer and catalyst -One part agent -Liquid metal -Gel -Additive to improve mechanical strength (CNT)	-Microcapsules (20–500 µm) -Small microcapsules (0.5–15 µm) -Elongated microcapsules (2 mm × 500 µm) -Hollow fibre -Microvascular network (2D and 3D)	-Ageing crack (slow) -Surface corrosion (slow) -Medium speed impact (<100 m/s) -High speed impact (100–500 m/s) -Hypervelocity impact (1–25 km/s)

Environment	Healing evaluation
-Air atmosphere pressure -Vacuum -Fixed cold temperature -Fixed high temperature -Cycled temperature -UV, X-ray, gamma radiation, etc.	-Slow three to four point bending test -Fast hitting mass in three point bending test -Compression after impact -Acoustic emission -Strain gauges -Optical fibre sensors -Infrared camera

Figure 7.1   Taxonomy of passive self-healing concepts. CNT, carbon nanotubes; 2D, two-dimensional; 3D, three dimensional

in 2007 [4]. It contained papers written by leading authorities in the field and covered a wide spectrum of materials, from polymers to metals and ceramics. Since that time, several reviews focusing on self-healing polymers have appeared in the literature [5–14]. Bergman and Wudl [5] described intrinsic healing in polymers and their mechanisms. Wool's contribution attempted to provide a general theory of damage and healing of polymers, drawing from the related field of polymer–polymer interfaces [6]. Wu *et al.* offered a primer on fracture mechanics and mechanisms of healing in polymeric systems [7]. The reviews by Kessler [8] and Yuan *et al.* [9] are more general and provide the context for ongoing research in the field. A recent *MRS Bulletin* was devoted to self-healing polymers [10–13], with contributions summarising self-healing chemistries [11]; polymers [12] and composite systems [13]. Recently, Trask *et al.* [14] provided a review of self-healing fibre-reinforced composites, while Ghosh [15] edited an excellent book dedicated to the design strategies and applications of self-healing materials.

Depending on the self-healing system that is being developed, healing agents based on monomers, epoxides or even dyes, and the catalyst are first integrated into microcapsules, hollow fibres or microvascular channels shaped reservoirs and then embedded into the polymeric system. Upon mechanical cracking, these reservoirs are ruptured and the reactive agent pours into the cracks by capillary force where it solidifies in the presence of the predispersed catalysts. This approach aims to stop the propagation of cracks. This process is often autonomous, without the need of an additional triggering processes [15].

Self-healing is usually considered as the recovery of mechanical strength through crack healing. However, there are other types of damage, for example, small pinholes that can be healed to ensure the proper performance of various materials. Self-healing polymers may be used to repair small punctures and pinholes. They show a great promise for mitigating potentially catastrophic damage from events such as micrometeoroid penetration and/or AO effects. Effective self-repair requires that these materials heal instantaneously following projectile penetration while retaining their structural integrity.

However, the main space environment parameters have to be taken into account to achieve an efficient self-healing concept. Figure 7.2 illustrates the various interactions between materials, structure and the space environment parameters.

It is imperative that all these elements are considered during the screening evaluation of the attribute on the potential candidates for the self-healing solution. For example, if the damage is contained within the laminate, that is, internal matrix cracks are induced by thermal loading, then a composite

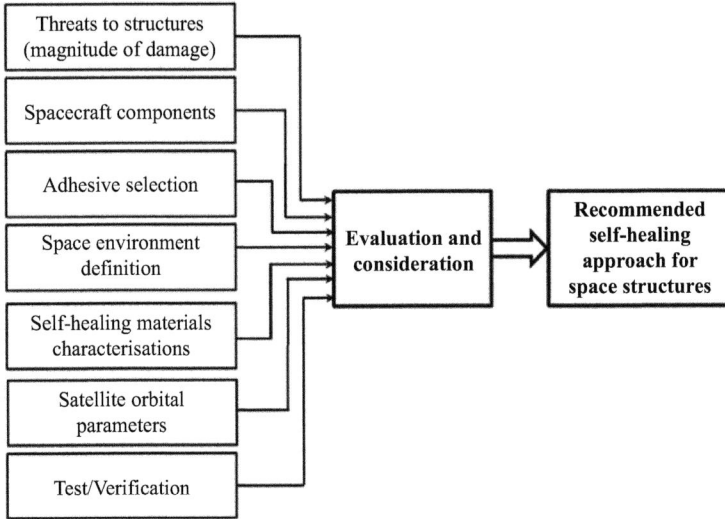

*Figure 7.2   Main technical considerations for the development of a
self-healing approach for space structures*

system will not be exposed to the AO and therefore, selecting a resin (or
composite) system and identifying the location of the self-healing plies
becomes more straightforward. By contrast, if the damage has arisen through
a micrometeoroid impact, the problem of AO and how this affects the self-
healing repair composite will have to be considered. Dimensional changes
because of the moisture desorption represent another critical challenge.

In summary, the properties of composite structures used in spacecraft,
the space environment, the damage modes induced in composites and the
space-adaptability of the different self-healing methods have to be determined before an effective solution may be implemented.

The possibility of self-healing composite materials has emerged from
concepts proposed by Dry [16]. Later on, these concepts were modified and
further developed by White *et al.* [17]. These systems are based on an
encapsulated healing agent that is embedded in a polymer. The tests performed confirmed that the self-healed composite material regained as much
as 90% of its original strength. Such a material is, therefore, able to sense
damage and initiate repair without the need of an external trigger or control.
Such a process is referred to as autonomous self-healing. The concept was
developed by a research group from the University of Urbana, Illinois [17].
It relies on a healing agent (crosslinking polymer) that forms a bond
between the two crack faces, and heals the structure (Figure 7.3). It has a
wide variety of potential applications for space structures.

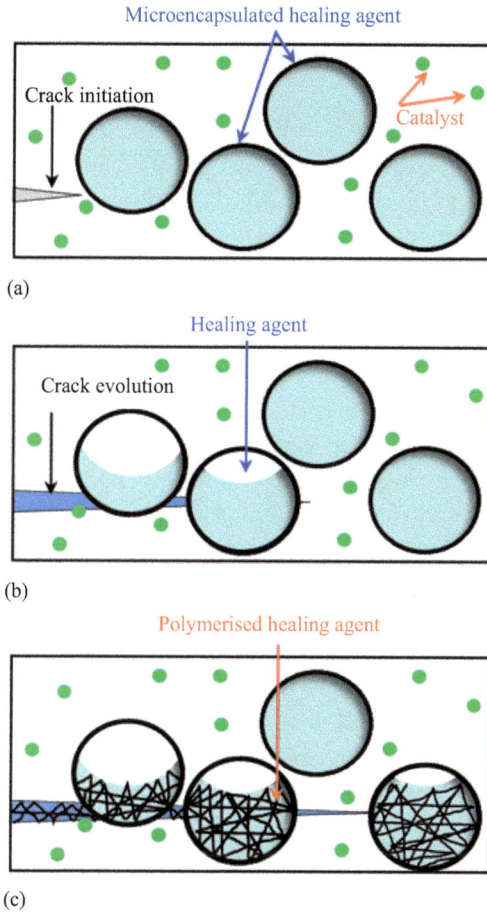

Figure 7.3   The self-healing process. (a) The healing agent, a monomer (e.g., dicyclopentadiene (DCPD)) is prepared and stored in microcapsules. The microcapsules and a catalyst are spread and embedded within the structure (matrix). (b) When a crack reaches a microcapsule, it causes it to rupture, which releases the monomer healing agent. (c) Self-healing is realised by polymerisation between the monomer and embedded catalyst

7.2   Approaches to space applications

This section presents an overview of several self-healing approaches that have been specifically developed for space applications. These newly developed technologies apply to composite structures, electrical wires, re-entry thermal resistant materials, propulsion tanks, human space suits, inflatable habitable structures and others.

7.2.1   Self-healing with microcapsules

Carbon fibre-reinforced polymer (CFRP) panels were prepared with DCPD embedded within microcapsules and hollow fibres. The Epon^{TM} 828 and Epon^{TM} 862 resins were used within the CFRP laminates structures. The CFRP were then subjected to indentation impact using a 1 kg block. The elongated microcapsules allowed a maximum amount of healing agents to be delivered to the damaged areas. However, such elongated microcapsules are difficult to make. The elongated capsules are obtained by deforming single drops, which are suspended in a liquid [18,19]. However, spherical microcapsules are much easier to make because of the interfacial tension between the dispersed phase and the continuous phase.

CFRP laminate specimens containing a healing agent between the two middle layers was submitted to a crack process. After the completion of the healing process, the samples were tested with a cycled load (up 250,000 cycles) using a three point flexure configuration. Half of the specimens were effectively healed. However, with continued cycling, the properties of the specimens indicated that the damage had returned to its original state. It was concluded that the healing achieved *via* the microencapsulation technique may be limited to the amount of healing agent available within the crack volume [20].

Composite cryotanks or composite overwrapped pressure vessels (COPV) offer a weight saving advantage when compared to the currently used metallic cryo-pumps. Because of its susceptibility to microcracks, the brittleness of the epoxy matrix used in the composite constitutes the main challenge. These cracks can either be caused by exposure to cryogenic conditions or by impact from outside sources. If the cracks are not prevented, the microcracks increase gas permeation and leakage. A reliable robustness of the COPV was obtained by using a combination of a modified resin and nanoparticle additives. The unique nanoparticles that were used have been surface-functionalised to be compatible with the resin [21].

A second epoxy with low viscosity resin was prepared using a surface modified nanomaterial additive. The low viscosity improved the fabrication and processing of the COPV. Preliminary results showed that the burst pressure of these new vessels was 20%–25% higher than that of the original. The amount of healing agent used was 20% of the total weight [21].

7.2.2   Self-healing with carbon nanotubes

There is a need to improve the repair patches that are used by the aerospace industry. Multiwalled carbon nanotube(s) (MWCNT)/epoxy and nickel-coated multiwalled carbon nanotube(s) (Ni-MWCNT)/epoxy systems were proposed as a possible solution. MWCNT and Ni-MWCNT increase the tensile strength and damping properties of the composite. The nickel coating is added to the MWCNT. The coating is thermally and electric conductive, magnetic and corrosion resistant. MWCNT and Ni-MWCNT were injected into carbon fibre composite repair patches *via* vacuum resin infusion. The MWCNT and the Ni-MWCNT did not modify the thermal stability of the epoxy system. The use of the probe sonicator to disperse them in the resin appears to have destabilised the hardener part of the epoxy. Work is progressing to optimise the integration of the MWCNT and Ni-MWCNT within the repair patches. The applications of this technology include spacecraft, commercial aircrafts, sports equipment and automobiles [22].

7.2.3   Self-healing of ceramics

Interest in developing SiC/SiC ceramic matrix composites (CMC) is because of their higher damage tolerance compared to monolithic ceramics. Current generation SiC CMC rely almost entirely on the SiC fibres to carry the load, leading to the premature cracking of the matrix. The innovative CMC concept is based on SiC fibre-reinforced SiC-Si_3N_4-silicide matrix composites with a composition formulated to match the coefficient of thermal expansion (CTE) of the fibres. The matrix composition converts any ingressed oxygen into low viscosity oxides or silicates, so that they can flow into the cracks by capillary action and seal them. For matrices containing $(Cr, Mo)_3Si$ silicides the expected amount of free silicon after melt infiltration is expected to be low, which would allow for use of the composites in applications at or above 1482 °C. Various silicides are proposed for testing such as magnesium, nickel, sodium, platinum, titanium and tungsten silicides [23].

7.2.4   Self-healing for re-entry vehicles

Ceramic powders such as SiC, ZrB_2 and Y_2O_3 are combined with allylhydride-polycarbosilane resin, and are mixed to form an adhesive paste. The material is then applied to the damaged area. This adhesive is capable of repairing damaged components of re-entry vehicles while in space. This is a novel approach because of its ability to be applied in a vacuum and microgravity environment. The material can be applied in space to repair damage that requires heat/oxidation protection upon re-entry to Earth's atmosphere. During the re-entry, the material is converted to a ceramic coating that provides thermal and oxidative stability to the repaired area, thus allowing the vehicle to pass safely from space into the upper atmosphere. The adhesive called nonoxide adhesive experimental (NOAX^{TM}) flew on all space shuttle missions from return to flight (STS-114) until the final flight (STS-135). It served as a crack repair material for the leading edges and nose cap of the vehicles [24,25].

7.2.5   Self-healing foams

A self-healing foam system can be incorporated between the rigid layers of a structure to repair the damage caused by punctures. Punctures of inflatable structures are a risk to all manned space missions. The inner foam system comprises separate encapsulated layers of two major components of the urethane foams: the polyol and the isocyanate. These components contain all the necessary catalysts, surfactants and blowing agents that are required for the self-healing. The two layers are assembled next to each other. In the event of a puncture, both layers rupture and bring the two components into contact. The foaming will rapidly seal the puncture. There are many spinoffs of this method for self-healing of fuel tanks on vehicles and aircraft [26].

7.2.6   Integrating sensing within self-healing structures

A joint team from NASA, Boeing and Sandia National Laboratories studied the sensing and self-healing of low crack formation in inflatable structures (Figure 7.4). They used self-healing polydimethylsiloxane (PDMS) elastomer matrix with embedded microcapsules. Separate microcapsules contained a

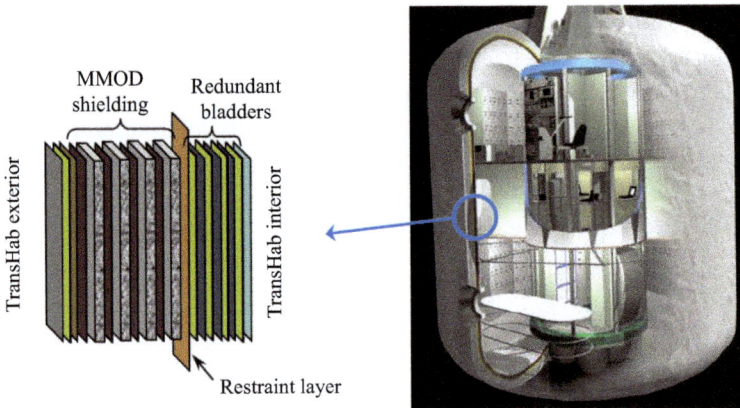

Figure 7.4   Cross-sectional schematic view of the Transit Habitat (TransHab) inflatable structure, highlighting the micrometeoroid and orbital debris shielding, restraint layer and redundant bladders. MMOD, micrometeorites and orbital debris. © 2006 NASA
[Reproduced, with permission, from [27]]

vinyl-terminated PDMS resin and a methyl hydrosiloxane copolymer. These two materials react *via* a Pt catalysed reaction. The Pt catalyst complexes are in solution with the vinyl-terminated resin and encapsulated together. PDMS was chosen as the first test material because of its high strain to failure (~200%), room temperature curing and the wide variety of adhesion promoting coupling agents. A commercial two-part PDMS system functioned as the matrix and provided the resin and initiator materials for the healing chemistry [27–29].

7.2.7   Self-healing paints

Microscopic nicks or pits on a surface develop during manufacturing or through wear and tear. This problem can be solved by the incorporation of a self-healing function into the coating. Several new concepts such as conductive polymers, nanoparticles and microcapsules are being developed to release corrosion-inhibiting ions at a defect site. Corrosion indicators, corrosion inhibitors, as well as self-healing agents have been encapsulated and dispersed into several paint systems to test the corrosion detection, inhibition and self-healing properties of the coating [30,31].

7.2.8   Self-healing of electrical insulation

A failure of wire insulation is considered to be a major problem for spacecraft [32,33]. This particular failure was the cause of several catastrophic accidents as follows:

- The Gemini 8 mission (1966): An electrical wiring short nearly resulted in the loss of the crew. The re-entry was in a region beyond the reach of a US tracking station. The US Department of Defense deployed 9,655 personnel, 96 aircraft and 16 ships to ensure the rescue of the crew;
- Shortly after the launch of STS-93 (July 1999), a primary and back-up main engine controller on separate engines dropped offline due to a short circuit of a 14 gauge Kapton® insulated wire;
- An arced wire in the in-flight entertainment network attributed to the loss of Swiss Air 111 (September 1998) by causing fire ignition and
- The loss of TWA 800 (July 1996) was attributed to a frayed wire in the centre tank area.

The proposed self-healing approach has the ability to repair damaged Kapton, Teflon or vinyl-type wire insulation. The self-repair must produce a flexible watertight seal over the damaged area. The self-repairing agent, incorporated in the insulator layer, is based on low-temperature melting polyamic acids and polyimides. The repair can be initiated by chemical, mechanical or electrical stimulus [33].

7.2.9 Foam layer surrounded conductor

This technology is the self-healing cable system that includes a conductor and an axially or radially compressible/expandable (C/E) foam layer that surrounds the conductor. The C/E foam layer is adapted to maintain its compressibility and expandability over a temperature range between −65 and +260 °C, and pressure range between high vacuum and the atmospheric ones. When a damaging force forms a breach in the protective jacket, a corresponding portion of the foam layer expands and covers the localised area [34,35].

7.2.10 Other self-healing products

In Sections 7.2.10.1–7.2.10.5, a brief review is given of commercially available products that may be used for partial self-healing of specific components and sub-systems.

7.2.10.1 Photosil™ graded layer

Photosil™ is a surface modification product, which alters the surface structure and chemistry of a polymer to protect the polymer materials. Under certain circumstances, the modified surface structure also has a self-healing capability. Photosil™ has been used to protect the surfaces of satellites and some elements of the space station robotic arm. It is used to protect the polymer-surface components of the Canadarm system. Photosil™ has also been used to treat polymer-surface components of the special purpose dextrous manipulator of the Canadarm on board the International Space Station (ISS) [36].

7.2.10.2 Self-healing using polyethylene-co-methacrylic acid

Nucrel® is the polyethylene-*co*-methacrylic acid (EMAA) copolymer. React-A-Seal and Surlyn® are EMAA ionomer-based materials. All three are manufactured by DuPont. Ionomers are polymers that contain ionic groups in relatively low concentrations along the polymer backbone. In the presence of oppositely charged ions these ionic groups form aggregates that lead to novel physical properties of the polymer (Figure 7.5). These ionomers are potentially applicable to self-healing of punctures in multilayer insulators (MLI) caused by small space debris and meteorites. The Surlyn polyethylene chains are interspersed with methacrylic acid to which ions are attached. Attractions between the ions form crosslinks within the material, a feature that gives specific properties [37]. Recently a preliminary test of commercial available Surlyn with hypervelocity pellets was introduced and showing a full repair of projectile puncture [38].

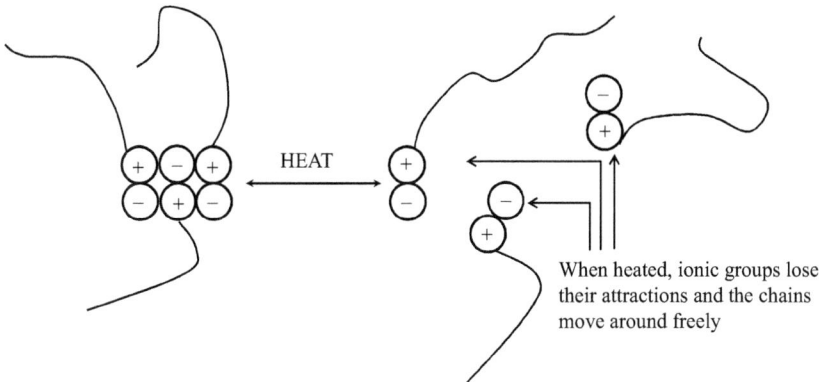

When heated, ionic groups lose their attractions and the chains move around freely

Figure 7.5   Schematic illustrating the ordering and disordering of an ionic aggregate with heat as the source of energy

7.2.10.3   Self-repairing shape-memory alloy ribbons

The goal of the self-sensing and self-repairing joint is to reduce the likelihood of failure because of self-loosening and to reduce the cost of the maintenance of critical bolted joints. The concept combines piezoelectric-based health-monitoring techniques with shape-memory alloy (SMA) actuators to restore tension in a loose bolt [39]. One of the main problems of the self-healing bolted joint is the triggering of the SMA actuators. The relatively large mass of the shape memory washer and its low resistance make resistive heating particularly difficult. Models were developed to assess the viability of resistive heating and provide an estimate for the power requirements for effective actuation. Modelling and experimental testing have shown that an external heater can be used to actuate an SMA actuator with conventional power sources. Making the SMA washer provides a convenient alternative to resistive heating, and aids the practical implementation of the concept of self-sensing, self-repairing joints. The SMA are used on the ISS to repair the joints/bolts.

7.2.10.4   Multifunctional copolymers

All living organisms utilise highly complex and specialised macro-molecules (e.g., deoxyribonucleic acid, proteins, peptides and sugars) to perform various biological tasks. Although most of these macromolecules have a complex chemical composition on a primary and secondary level, it is often their tertiary structure, or supra-molecular organisation, that leads to materials with unusual chemical and mechanical properties (e.g., those found in spider silk and collagen) [39]. On a primary and secondary molecular level one could access linear, branched or grafted polymers that are amorphous, crystalline or semicrystalline. Most polymers were synthesised with one specific function, and were used as adhesive, film or fibre.

A special technique, known as 'living anionic polymerisation', has provided the organic polymer chemist with a tool to go beyond those limitations. Now it is possible to combine polymers with different chemical and physical properties in one polymer backbone. As in nature, polymers can be synthesised in a way that allows materials to be obtained with molecular control.

7.2.10.5   Self-healing composites with electromagnetic functionality

Electromagnetic effective media is usually using arrays of electrically conducting elements (such as metallic wires) that are serving as antennas. The incorporation of these conducting elements into fibre-reinforced polymer and/or ceramic-based composites has found to provide many advantages, including the multifunctionality of the self-healing composite materials and the ability to the thermal transport.

In addition to desired structural properties, these electromagnetic media can provide controlled response to electromagnetic radiation such as radio frequency (RF) signals, radar and/or infrared (IR) radiation (Figure 7.6).

Braided coil media	**Embedded in a crack-heating polymer matrix**	**Final product laminated to form a structural composite**
	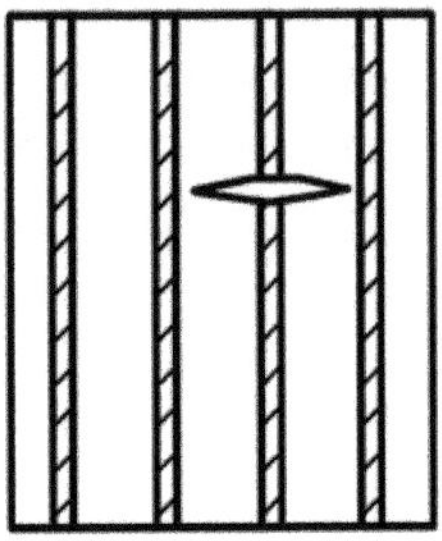	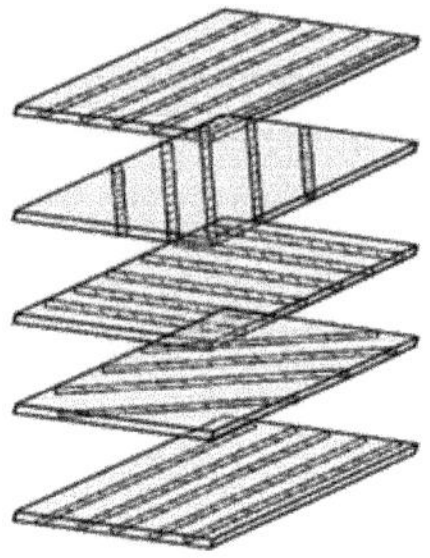
- Structural fibres (Kevlar, glass) - Strength and toughness - Coil media (copper wire) - Electromagnetic (EM) functionality - Heating elements - Center core (fibres, ceramics, rod) - Reinforcement	Matrix may be repaired with heat (Lower or negative coefficient of thermal expansion of the core constrains matrix to close the crack faces)	- Structural integrity - Tuned EM transparency - Sensing and self-healing - Thermal management

Figure 7.6   Self-healing process with embedded electrically conductive wire

Application of heat allows fractured bonds to reform and thus repair the damaged interface. Since the repair mechanism is not automatically activated it may not be considered an autonomous healing material. It constitutes a self-healing function, particularly when the healing agent (heat source) is integrated into the material in the form of conductive metal wires [40].

7.3 Materials ageing and degradation in space

A major challenge for space missions is the materials' degradation and failure over time, because of natural ageing, and the extreme conditions as well as the external impacts that characterise the space environment. Materials suffer mainly from degradation and erosion under exposure to AO, vacuum ultraviolet radiation (VUV), temperature swing, vacuum outgassing and space debris. This section briefly reviews mechanical ageing and the main degradation mechanisms.

7.3.1 Mechanical ageing

A natural mechanical degradation occurs through the ageing of materials and equipment. The common evolution of failure rate over time, for example, the development of cracks in a structure, follows a well-known graph which has a bathtub shape (Figure 7.7). The curve consists of three parts. The first (Phase I, on the left-hand side) is the failure rate just after manufacturing. It is called the initial failure rate. Phase I is also called 'infant mortality' since it is similar to the high rate infant mortality observed in human and animals. Phase I shows that, if there is any hidden defect in a manufactured unit, the defects will cause a premature failure. The second part (Phase II) consists of an almost stable evolution of the failure and has rather a small rate over the major part of the life time. Phase III (the third part, just before the end of the lifetime of the product) presents a high failure rate that increases with time.

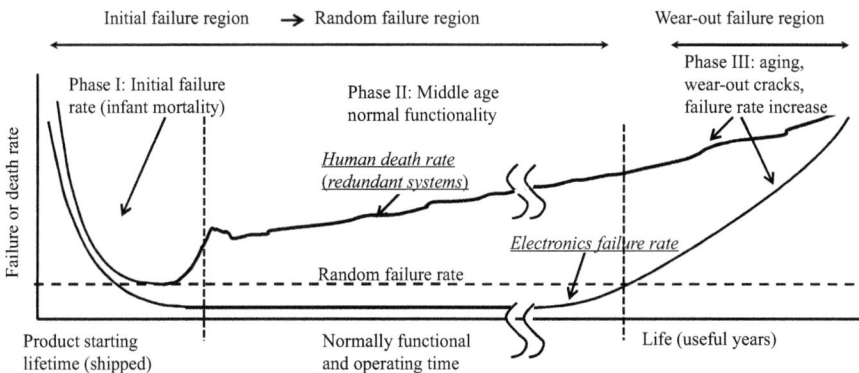

Figure 7.7 Common failure rate as function of time after manufacturing

The graph is cut in the middle to show Phases I and III better. The ageing effects are observed in the third part of the curve (on the right-hand side, i.e., wear out failure rate). In this phase, the stress begins to be seen through the development of cracks and fissures. Phase III may be divided into three stages. In the first stage, the fissures or cracks develop very fast, then a second stage follows where the speed of crack progress stays constant for a while before the cracks get too large and pass to the third stage, where the large cracks cause a complete failure.

The human death rate follows somehow a different shape with the minimum of the curve staying only a relatively short period (the U shape, from approximately 5–20 years of age). The death rate continues increasing slowly with time up to the end of life. This curve was not very well understood until the time, when the failures of computers with redundant processors were observed. The lifetime of computers has manifested similar characteristics to that of the human lifetime. This has allowed the realisation that the human death rate curve is influenced by the redundancy of human organs, such as lungs, kidneys, heart and so on.

Space systems include redundancy of the important components or those that may fail before the end of expected lifetime. According to Gavrilov and Gavrilova [41], the failure rate of space systems as a function of ageing would follow a curved shape similar to human death rates. There is no further study of these two lifetime characteristics.

To ensure functionality of spacecraft for a long time, the screening of materials and components greatly reduces the infant mortality. This screening is comparable to the common 'burn-in' in semiconductor manufacturing, where the components are tested for a certain period of time for their survival of the infant mortality phase. The screening for space applications consists of testing selected samples of materials and components. If there is a risk of failure due to a hidden defect in the unit, it will show up during the screening burn-in test.

Table 7.4 summarises the main degradation mechanisms in space and their effect on equipment and structures.

Self-healing is needed to improve the lifetime of the structure and to repair punctures and delamination caused by meteorites. The LEO and the ISS have typical orbits between 400 and 800 km. The conditions for geostationary orbit (GEO) are similar but with higher exposure to radiation flux. Table 7.5 shows an example of the LEO space environment for a usual solar year. If a solar flare occurs, then the level of radiation, including UV visible and IR, can be two orders of magnitude higher. At GEO orbits, two main differences are the disappearance of AO and a larger range of temperature variations.

Table 7.4   Main degradation mechanisms in space and their effect on equipment and structures

Degradation mechanism	Effects
Micrometeorites and debris	Pinholes and punctures
Temperature swings from −150 to +150 °C	Deformation and cracks developed by thermal ageing
High vacuum effect	Outgassing of polymers
Ageing structures	Deformation of structural parts under load and stress – development of cracks
Atomic oxygen	Erosion and oxidation
Radiation: protons, gamma and electron	Erosion, oxidation, mechanical degradation and chemical change
Shock and vibration during the launch	Development of mechanical stresses and cracks

Table 7.5   Space environment conditions for LEO orbit for a typical solar year

Parameter	LEO	Effects
Vacuum	10^{-3} Pa	Outgassing can lead to loss of volatiles
Temperature cycles and swings	−150 to +150 °C: outside-LEO −200 to +200 °C: outside-GEO 40 to +80 °C: inside spacecraft	Thermal stress on materials – the thermal heat generated on the sun-facing side can be used to assist the healing process
X-rays	10^{-3} W/cm^2, CuKα: 60 keV	Erosion and thermal stress, low energy may be useful too
VUV radiation	0.75 μW/cm^2, λ (100–150 nm) 11 μW/cm^2, λ (200–300 nm)	Degrades some polymers, beneficial for healing/curing
Protons and ions	15 krad/y, 1.1×10^{10} proton/cm^2/y	Erosion/oxidation of some materials, could be used to enhance polymerisation
Electron-based	1.5×10^{13} electrons/cm^2/y	Minor erosion effects, electrostatic discharge (ESD) can cause local damage and burn holes
Atomic oxygen (for LEO orbit)	Total fluence: 9×10^{17} particles/cm^2 Flux 10^{12} to 10^{14} atom/cm^2/min	Erosion of polymers, protect using SiO$_x$, VO$_2$ coatings and so on

Table 7.6 illustrates a summary prepared by NASA of space environment phenomena, their programmatic issues, the models and databases used to evaluate their effects and their effects on materials and their optical properties [42].

The damage either occurs very quickly as in punctures caused by meteorites or very slowly as in cracks developed as the result of ageing.

Table 7.6 *Summary of space environment phenomena, their related affected items and their effect on materials and their optical properties*

Space impacts	Definition	Affected items	Materials and optical properties
Neutral thermosphere (upper atmosphere region)	Atmospheric density, density variations, atmospheric composition AO and winds	Guidance, navigation and control (GN&C), system design, material degradation/surface erosion (AO fluences), drag/decay, spacecraft (S/C) lifetime, collision avoidance, sensor pointing, experiment design, orbital positional errors and tracking loss	Material selection, material degradation, S/C glow and interference with sensors
Thermal environment	Solar radiation (albedo and outgoing long-wave radiation (OLR) variations), radiative transfer and atmospheric transmittance	Passive and active thermal control system design, radiator sizing/material selection, power allocation and solar array design	Material selection and influences on optical design
Plasma	Ionospheric plasma, auroral plasma and magnetospheric plasma	Electromagnetic interference (EMI), S/C power system design, material determination, S/C heating and S/C charging/arcing	Arcing, sputtering, contamination effects on surface properties and change or degradation in surface optical properties
Meteorites and orbital debris	Meteorites and orbital debris (M/OD) flux, size distribution, mass distribution, velocity distribution and directionality	Collision avoidance, crew survivability, secondary ejecta effects, structural design/shielding and materials/solar panel deterioration	Degradation of surface optical properties
Solar environment	Solar physics and dynamics, solar activity predictions, solar/geomagnetic indices, solar constant and solar spectrum	Solar prediction, lifetime/drag assessments, re-entry loads/heating, input for other models and contingency operations	Solar UV exposure needed for material selection and necessary data for optical designs
Ionising radiation	Trapped proton/electron radiation, galactic cosmic rays and solar particle events	Radiation levels, electronics/parts dose, electronics/single event upset, materials dose levels and human dose levels	Darkening of windows and fibre optics
Magnetic field	Natural magnetic field	Induced currents in large structures, locating South Atlantic anomaly (zone where there is an increase of radiation flux close to the Earth) and location of radiation belts	
Gravitational field	Natural gravitational field	Orbital mechanics/tracking	
Mesosphere (50–80 km from Earth) S/C: Spacecraft	Atmospheric density, density variations and winds	Re-entry, materials selection and tether experiment design	Degradation of materials due to atmospheric interactions

Depending on their exposure to space damage, the space structures can be divided into three categories:

- *Outer spacecraft structures*: These structures are exposed to the outer space environment with high temperature swings, for example, from −150 to +150 °C. The failure is developed by thermal and mechanical stresses. These structures have a protective coating and are less affected by AO.
- *Internal structures*: The variation of their temperature is much less (−40 to +70 °C) and they are protected against space debris and AO. They cause ageing and cracks.
- *Outer spacecraft surfaces*: They are exposed to erosion through exposure to AO. The damage is in form of pinholes and punctures caused by the debris and small meteorites. Solar cells, MLI blankets and sunshields are examples of such surfaces.

7.3.2   Meteorites and small debris

Space activities and fragmentations happening on Earth orbit have created a large amount of man-made space debris. Together with natural objects (meteorites) they contribute to the Earth particulate environment. As of December 2013, NASA had counted about 17,000 objects in Earth orbit, which were tracked and registered by the US Space Command. The small size un-catalogued population as modelled by the ESA MASTER Model consists of more than 400,000 objects larger than 1 cm, about 1.8×10^8 objects larger than 1 mm and more than 1.2×10^{11} objects larger than 0.1 mm. In the smaller size category, paint flakes and residues from solid rocket motor firings are known to contribute to the debris. Presently, the man-made debris environment in most LEO regions is assumed to be larger than the meteorite contribution, except for the size category of around 0.1 mm diameter.

Meteorites and small debris are one of the main causes of the degradation of the coatings and external surfaces, mainly the MLI that envelopes the equipment to protect them against thermal fluctuation or AO. They are one of the main problems of the ISS outer surfaces. Figure 7.8 illustrates the average number of small meteorites in LEO orbits (400 km). The ISS and the Hubble Space Telescope are two examples of space structures with a large number of cracks observed in their MLI which are caused by the meteorites. In the first Hubble service mission, launched in December 1993 (SM1), some obvious damage was observed but only on the anti-solar side. However, in the second service mission (SM2) that went up in February 1997, many more effects were observed:

- More than 100 obvious cracks,
- Severe cracking on both the solar and anti-solar side and
- Some cracks were curled.

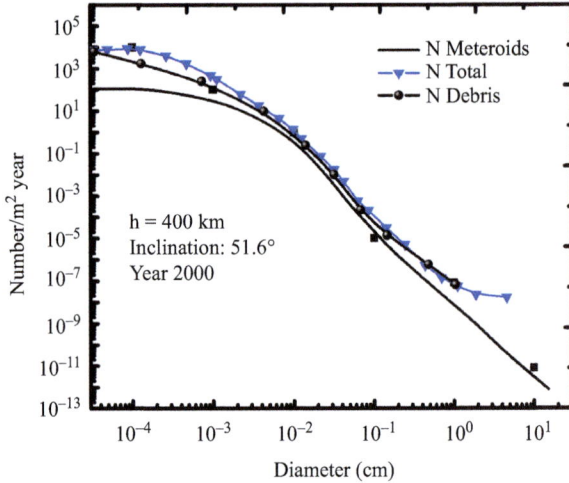

*Figure 7.8   Number of meteorites and small debris (per year/m²) in LEO
Orbit and ISS. © 1999 Oxford Brookes University-ESA-ESTEC
[Reproduced, with permission, from [43]]*

Space debris coming from human devices sent into space are present mainly in the LEO below 2,000 km, and around the altitude of the GEO. Meteorites, which are a natural phenomenon, are found everywhere in space:

- Impact effects from meteorites and debris are similar;
- Average impact velocities in the LEO are 10 km/s for space debris and 20 km/s for meteorites:
  - Orbital debris: V_{impact} = 2–15 km/s and
  - Natural micrometeorites: V_{impact} = 2–72 km/s;
- The average material density of meteorites is lower than that of the space debris;
- In the LEO, meteorites dominate in debris of sizes between 5 and 500 μm (0.5 mm) and
- Space debris dominates for larger sizes.

Micrometeorites are small particles from an asteroid or comet orbiting the Sun that survive their passage through the Earth's atmosphere and impact the Earth or the satellite surface [43].

Hypervelocity impact events may modify the original chemical composition of an impactor, fractionating volatile from refractory elements. Thus, micrometeorite residues may not necessarily retain the stoichiometric chemical signature of their parent mineral; in such a case analytical results are not compatible with those of mineral standards. Notwithstanding such difficulties, energy dispersive (EDS) spectra and X-ray elemental maps of residues that contain the following elements can be used as indicators of micrometeorite origin [43]:

- Mg + Si + Fe (mafic silicates, e.g., olivine or orthopyroxene);
- Mg, Ca, Na, Fe, Al, Ti + Si (clinopyroxene);
- Fe + S (iron sulphides);
- Fe + Ni (minor or trace) + S (Fe-Ni sulphides);
- Fe + Ni concentration at meteoritic levels (metal);
- Si + C (silicon carbide);
- Fe, Mg, Al + Si (phyllosilicates, e.g., serpentine) ;
- Ca, C, O (calcite) and
- Cl, Cr, K and P have also been individually identified in meteoritic samples and therefore, may be indicative of a micrometeorite origin under some circumstances.

Because of the complexity of any original micrometeorite (poly-mineralic composition), it is possible that a single impactor could be any one of the many combinations in the previous list.

The remnants of space debris material may be identified from the chemistry of their residue by using EDS spectra and X-ray elemental maps that contain the following elements [43]:

- Mainly Ti + possible minor traces of C, N, O, Zn (paint fragment);
- Mainly Fe + variable Cr, Mn + possible traces of Ni (specialised steels);
- Mainly Al + minor traces of Cl, O, C (rocket propellant);
- Mainly Sn + Cu (computer or electronic components) and
- Mg, Si, Ce, Ca, K, Al, Zn (glass impactor, possibly from other solar cells).

The presence of the Ti/Al layer within the solar cell complicates the identification of artificial impacts since Ti has been traditionally used as an indicator of paint fragment impact. In the Hubble Space Telescope, solar cells containing Ti, Al and Ag were attributed to artificial debris particles, such as paint fragments. Thus, Ti is probably a good indicator of paint fragments. When observed along with Al and Ag, it is more likely to represent a melt from the host solar cell.

The classification of impact residues as either space debris or as micrometeorite in origin is extremely complex, and often it is not possible to give a totally unambiguous answer. For example, although it is highly likely that a residue composed of Al and O is the remnant of solid rocket motor debris (Al_2O_3), it could conceivably also be corundum (Al_2O_3) which has been identified in primitive meteorites, although it is extremely rare.

Apart from the classification criteria given for residual material of either micrometeorite or space debris origin, there is a strong possibility that spacecraft and satellite surfaces may also be subject to contamination. There are several different possible sources of contamination arising from laboratory handling, to ground exposure, or to the space environment itself, whereby contaminants are effectively encountered at low velocity and thus are only loosely bound.

Hypervelocity impacts create a shock wave in the material and lead to very high pressures (>100 GPa) and temperatures higher than 9,727 °C. Further information is given, for example, in the *ESA Space Debris Mitigation Handbook* [44]:

- The impact process lasts only a few microseconds;
- The impactor and the target material are fragmented, often molten and/or vaporised, depending on the impact velocity and materials;
- Most of the impact energy ends up in the ejecta (i.e., ejected mass);
- The ejecta can be much larger than the mass of the impactor and
- A small fraction (less than 1%) of the ejected material is ionised. This latter phenomenon is a function of the impactor velocity.

On the other hand, collision damage depends on

- Kinetic energy of the particle (speed);
- Design of the spacecraft (bumpers, external exposure points) and
- Collision geometry (especially the angle of collision).

The impact ranges are about

- 1 cm (medium) at 10 km/s: this can fatally damage a spacecraft and
- 1 mm and less: which erodes the thermal surfaces, damage the optics and puncture the fuel lines.

The near space environment is actually very polluted by significant traces of recent human space history. All the space vehicles that have left the Earth have participated in this growth of collision risks in space. The population of space debris is composed of a very large variety of parts from the smallest (less than a millimetre) up to entire vehicles (up to several tons for lost spacecraft).

Figure 7.9 shows the spread of collision debris orbital planes [45]. Table 7.7 summarises the size distribution of space debris and micrometeorites.

Numerous discussions on the response of metallic structures due to the hypervelocity impact (up to 7 km/s) have been published. The damage of composite panels produced by low velocity impact has been studied by several authors [46–50]. For metals, the impact produced a dent or localised

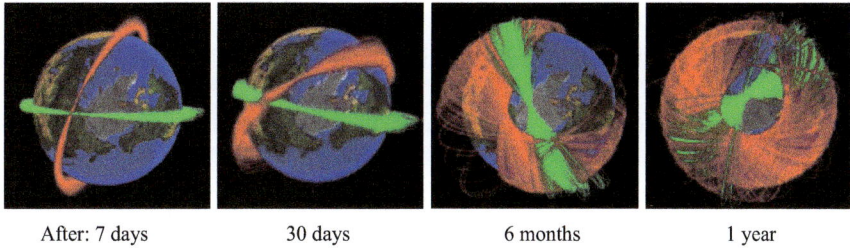

| After: 7 days | 30 days | 6 months | 1 year |

Figure 7.9 Spread of collision debris orbital planes. © 2009 NASA
[Reproduced, with permission, from [45]]

Table 7.7 Summary of the size distribution of space debris and
micrometeorites in the LEO

Category (and origin)	Size	Numbers in orbit	Probability of collision (and effects)
Large: Debris (satellites, rocket bodies and fragmentation material)	>10 cm	1 to 5 × 10^4 (low) (17,800 in 2001)	1/1000 (collision results in total breakup and loss of capability)
Medium: Fragmentation debris, explosion debris and leaking coolant	1 mm − 10 cm	1 to 5 × 10^6 (medium) (0.5 × 10^6 in 2001)	1/100 (collision could cause significant damage and possible failure)
Small: aluminium oxide particles, paint chips, exhaust products, bolts, caps and meteoric dust	<1 mm	>10^{10} (high) (3 × 10^8 in 2001)	Almost 1/1 (collision should cause insignificant damage)

plastic deformation at the impact area. The damage of composite panels was mainly due to fibre and matrix breakage and layer delamination [51].

Schonberg [52] presented an overview of the trend of research of the composite structural systems in protecting the Earth orbiting spacecraft against hypervelocity impact damage. According to the findings of various studies, the effect and extent of damage of MMOD impact on a spacecraft structure depends on many independent as well as inter-related factors. These include size, shape, density, composition and relative speed of the impactor, impact energy, material and structural properties of the target, thickness of the laminate, lay up sequence, the angle of impact and so on. The principal mechanism for failure of the composite structure due to this type of loading includes the fibre breakage and fibrillation, fibre/matrix debonding, delamination, matrix deformation, spallation and impact cratering [53]. The extent of damage in most studies was characterised by the dimension of the crater hole produced by the hypervelocity impact, penetration depth, total area of damage, spalled back and front surface, secondary ejecta plumes and so on [51,54–56]. Yew and Kendrick [51] observed multiple breakage and delamination of the laminate and matrix material in a graphite fibre composite panel under hypervelocity impact. They correlated the process of perforation and propagation of damage (e.g., the shear plugging and ejection at the rear face) in the plate with the shock wave motion produced by the impact. The methods of analysing the impacted specimen found in several studies can be divided into three categories:

- The nondestructive postimpact test, which includes the observation of the damaged sample using digital photograph, ultrasonic scanning and X-ray tomography [57];
- Destructive postimpact tests such as tensile tests, sectional viewing under a microscope [51], which needs careful surface preparation and substantial postimpact handling of the specimen and
- The observation of material response *in situ*. Chambers and co-workers [58] used fibre optic sensors embedded inside the CFRP laminate to monitor the residual strain response during the low (1.3 m/s) and high (225 m/s) velocity impacts on the sample.

7.3.3   Atomic oxygen effects

The LEO AO environment poses unique durability problems to spacecraft. Its energy and flux along with a highly chemical activity can oxidise most materials that are typically used for spacecraft manufacturing. The results of the oxidation can lead to component structural failure, loss of thermal control or contamination depending upon the composition and function of the materials.

AO is formed in the LEO environment by photodissociation of diatomic oxygen. Short wavelength (<243 nm) solar radiation has sufficient energy to break the 5.12 eV O_2 diatomic bond in an environment where the mean free path is sufficiently long (100 m) so that the probability of re-association or the formation of O_3 is small [59]. The reaction of AO with spacecraft materials has been a significant problem to LEO spacecraft designers. The spacecraft impacts the AO resident in LEO with sufficient energy (about 7 km/s) to induce chemical reactions.

AO can break hydrocarbon polymer bonds, and can react with carbon and many metals to form oxygen bonds with the atoms on the surface being exposed. For most polymers hydrogen abstraction, oxygen addition or oxygen insertion can occur. With continued AO exposure, all oxygen interaction pathways eventually lead to volatile oxidation products accompanied by the gradual erosion of hydrocarbon materials. Surfaces of polymers exposed to

AO also develop an increase in oxygen content, causing oxidation and thinning of the polymers due to loss of volatile oxidation products. AO can also oxidise silicones and silicone contamination to produce nonvolatile silica deposits. Such contaminants are present on most LEO missions and can be a threat to the performance of optical surfaces.

Some polymeric materials such as Kapton were proposed for use in many inflatable structures in space, however, they are severely eroded by AO, and need to be hardened against it.

Most of the methods to reduce the AO effect are based on depositing a thin protective film or implanting oxide molecules within the polymers. Various tests on pulsed vapour deposition (PVD) coatings of Si, SiO_x, SiN and SiON have shown that these coatings provide good protection of the underlying polymers [60]. Other depositions which perform well were demonstrated using laser ablation, plasma and electron cyclotron resonance (ECR)-PVD techniques [60]. Although their deposition was successful, there are still some severe constraints. The coefficient of thermal expansion (CTE) of the coatings mentioned differ from those of polymers, which can result in cracks developing during thermal cycling or if AO penetrates through to the substrate. Some groups succeeded in solving this problem by including interfaces with good adherence to the polymer and to SiO_x [60].

A different option is to use siloxane composites, which are converted to SiO_2 by AO interaction [61–63]. Such surfaces are stable in a vacuum and prevent further oxidation occurring. Siloxanes can, therefore, be exploited first as coatings but also as additives in the polymeric matrix. One of the alternative approach to create such a protective coating is based, for example, on the UV/ozone treatment of polymers embedding PDMS additives (in low concentrations, typically in the range of 0.1–2.0 wt%), and was found to produce successfully oxidised layers [30] that protect efficiently the underlying material from further degradation [59].

Another effective approach is to incorporate polyhedral oligomeric silsesquioxane (POSS) into a polyimide (Kapton) matrix by copolymerisation, which distributes Si and O in the polymer matrix on the nanolevel. During exposure of POSS polyimide to AO, organic material is degraded and a silica passivation layer is formed [61,62]. A few samples of POSS, with various percentages of Si_8O_{11} and Si_8O_{12}, were exposed to AO and UV in two space mission: Materials International Space Station Experiments (MISSE), MISSE-5 (2005-2006) and MISSE-6 (2008-2009).

Alternative technologies such as Photosil [36] or the Implantox process [64] can also be used to protect spacecraft surfaces. However, such treatments have to be applied on the ground with substantial costs, and the solar absorbance of the material can be significantly altered.

In summary, the proposed methods to reduce the effects of the AO are as follows:

- Covering the surface with a coating with SiO_2 or (Si_nO_m). SiO_2 do not exhibit good, reliable adhesion when directly applied to plastics because of the interfacial stress. Considerable effort has devoted to develop special interfaces that facilitate the adhesion of high-quality SiO_2, to various polymers and resins including Kapton, polycarbonate, Teflon and even fibre glass in an integrated thin film structure;
- Use of siloxane composites, which are converted to SiO_2 by AO interaction. Siloxane is easier to adhere to some surfaces than SiO_2;
- Selection of an appropriate method of deposition (plasma and laser ablation);
- Surface texturing or processing to tailor the surface characteristics of Kapton and Teflon for better adherence to SiO_2;
- Stressing the polymer, to improve its resistance;
- Implanting the polymer with a small percentage of Si_nO_m and
- Hybrid implanting and depositing of SiO_x [62,65].

Figures 7.10–7.12 and Table 7.8 illustrate the original effect of AO on Teflon and Kapton and the reduction of the effect after adding SiO_2 over a special interface to ensure the adhesion of SiO_2 to Teflon and Kapton.

Once formed, the oxide layer acts as a protective layer against further AO attack [63]. An exception is silver that is easily oxidised. The silver foils, used as interconnectors in solar cells, are protected by plating with gold.

No degradation is reported on glasses and ceramics, which are based on oxides that are saturated with oxygen.

Figure 7.13 gives examples of the erosion yields of various materials used in space, assembled from different sources.

7.3.4 Vacuum effect

The space vacuum environment adds more challenges. The specific reactions related to the self-healing system and the components that might outgas in vacuum are as follows:

- The resin structure: the resin is in solid state, its outgassing is low, but at high temperatures the resin starts to melt and the outgassing will increase and can limit the reaction;
- The Grubbs' catalyst used in the self-healing reaction is in a solid nanopowder form, and is based on a noble metal (ruthenium) and so the probability of outgassing occurring is very low;

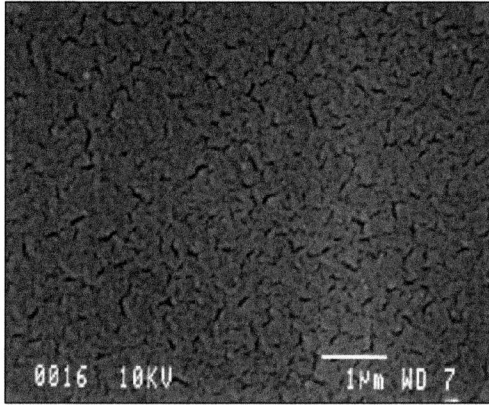

(a)

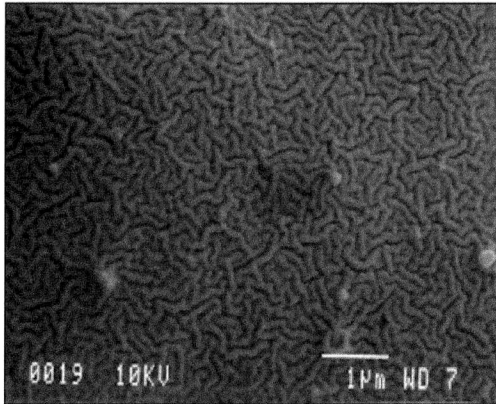

(b)

Figure 7.10 SEM micrographs showing the morphology of a (a) 5 mil thick
bare type A Teflon FEP and (b) 1,500 Å thick SiO$_2$/(interface)/
Teflon FEP, as deposited at 450 W by ECR-CVD. SEM, scanning
electron microscopy; CVD, chemical vapour deposition;
FEP, fluorinated ethylene-propylene. © 1998 ICPMSE-4
[Reproduced, with permission, from [60]]

- The shell of the microcapsules is made from poly-urea-formaldehyde. It
  may outgas in space vacuum, unless it is embedded within a hermetic
  host material such as a solid phase resin and
- The healing agent is a liquid monomer. It may outgas fast when the
  microcapsule shell is broken during the development of a crack.

(a)

(b)

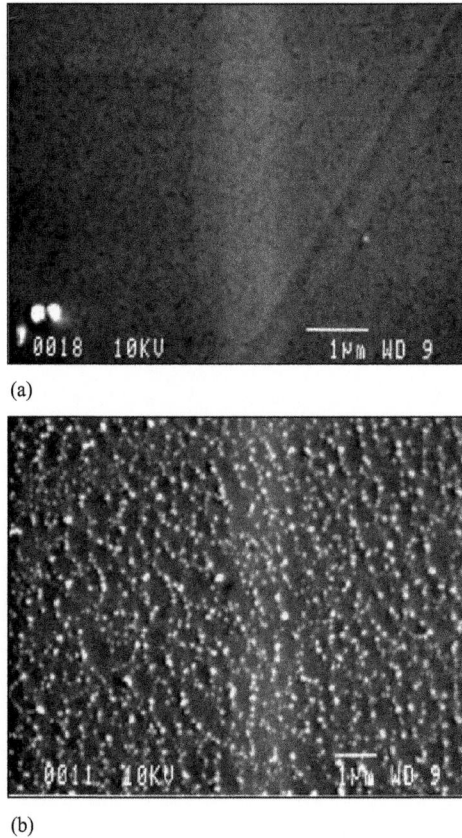

Figure 7.11   SEM micrographs of (a) (2,000 Å SiO₂)/(proprietary
interface)/Kapton film 'as deposited' onto Kapton at 350 W
by ECR-CVD after exposure to a VUV-AO source, and (b)
bare Kapton after a similar AO exposure. © 1998 ICPMSE-4
[Reproduced, with permission, from [60]]

In outer space, once a crack starts to develop, it breaks the microcapsules
and the monomer flows through the broken shell (Figure 7.14), and then
simultaneously:

- The monomer starts to react with the catalyst to form a stable polymer
  (healing) and
- At the same time some parts of the monomer start to evaporate.

There is competition between the healing chemical reaction (polymerisation) and the evaporation. In order to permit the healing to happen in
the space environment, the polymerisation and curing processes should be
faster than the evaporation process.

(a)

(b)

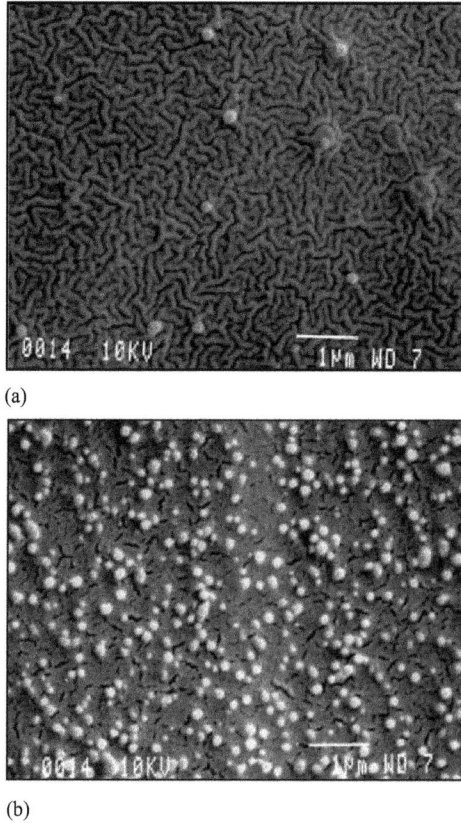

Figure 7.12 *SEM micrographs of (a) 1,500 Å thick SiO₂/ proprietary*
interface film on 5 mil Teflon FEP as deposited at 450 W by
ECR-CVD after exposure to VUV-AO source and (b) bare
Teflon FEP after similar exposure to AO source. © 1998
ICPMSE-4
[Reproduced, with permission, from [60]]

Table 7.8   Summary of the test of VUV-AO

Sample	Erosion depth (µm)
A piece of Kapton	10.8
SiO₂/Kapton	<0.01
Bare aluminised Teflon	24−25
SiO₂ aluminised Teflon	<0.01
Bare 5 ml Teflon FEP	15
SiO₂/Teflon FEP	<0.01

Reproduced, with permission, from [28], ©Marshall Space Flight.

(a)

(b)

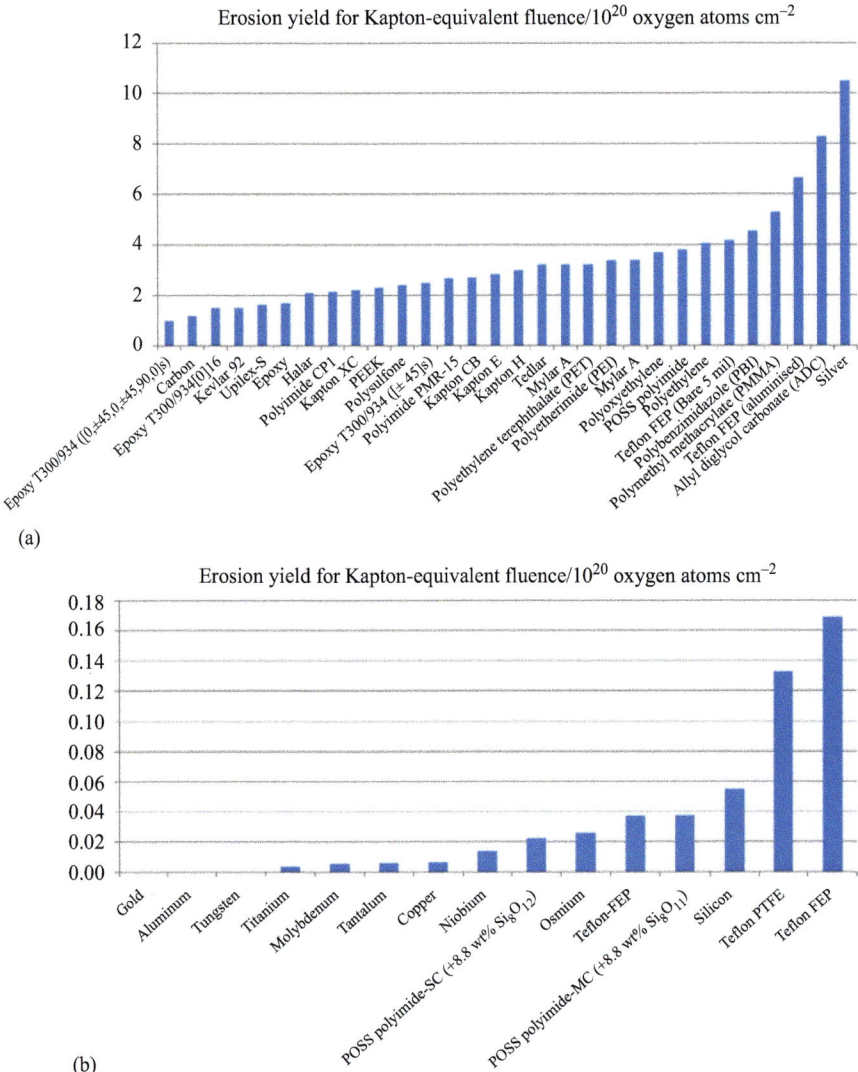

Figure 7.13   *Erosion yield of different materials, from different references,*
normalised to Kapton: (a) medium and high erosion yield
materials and (b) low yield materials (low erosion). ADC,
allyl diglycol carbonate; PBI, polybenzimidazole; PEEK,
polyether ether ketone; PEI, polyetherimide; PET,
polyethylene terephthalate; PMMA, polymethyl methacrylate;
PTFE, polytetrafluoroethylene; MC, main-chain; SC,
side-chain

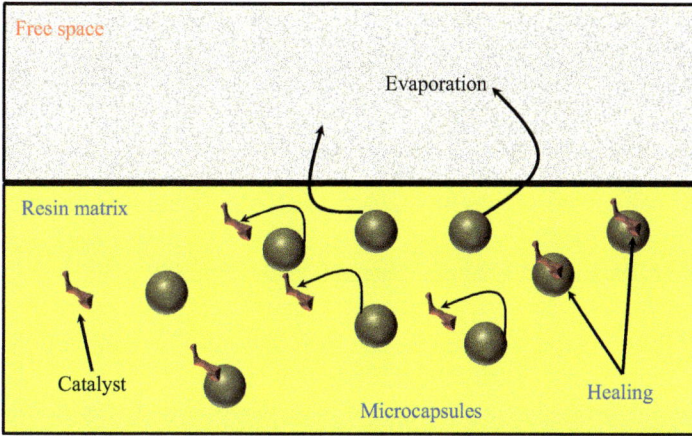

Figure 7.14 Competition between healing and evaporation phenomena

The healing reaction using monomers (e.g., DCPD or 5-ethylidene-2-norbornene (ENB)) is a polymerisation reaction similar to resin curing. Resin curing is of particular interest when making inflatable space structures rigid [66,67]. The polymerisation requires an initiation of the reaction, which can be induced by temperature or radiation. For self-healing, the initiation is triggered by the catalyst. The similarity between polymerisation and resin curing may be used to predict healing reaction in space.

Resin curing in space is becoming very important because of the use of polymeric material in the main inflatable structures. These structures can be folded during the launch and transportation in space up to orbit or at their final location. They are then unfolded and kept in their final shape after curing the epoxy holding the structures. The resin curing can be triggered and maintained by the solar UV radiation.

In its simplest form, the rate of evaporation in a vacuum is described by the Langmuir formula:

$$W\left(g/\left(cm^2 \times sec\right)\right) = A \times P \times (M/T)^{0.5} \tag{7.1}$$

where A is the constant; W is the rate of evaporation ($g/(cm^2 \cdot s)$); M is the molar mass of vapour (g/mol); T is the temperature (K) and P is the equilibrium vapour pressure of fraction from the Clausius–Clapeyron equation.

Materials with a low vapour pressure, such as metals, will have low evaporation rates. In the nineteenth century Heinrich Hertz demonstrated experimentally the linear relationship between the evaporation rate and the pressure. This relationship is consistent with the kinetic theory of gases, in which the impingement rates, and thus, the evaporation, are proportional to the pressure.

The evaporation rate is inversely proportional to $(T)^{0.5}$. This factor originates from the molecular energy (velocity) distribution, provided by the classical kinetic theory of gases [68]:

$$\partial N(v)/\partial N = C(M/T)^{0.5}v^2\exp(-mv^2/2kT)\partial v \tag{7.2}$$

where N is the number of gas molecules; v is the molecule gas velocity (m/s); M is the molar mass of vapour (g/mol); T is the temperature (K); k is the Boltzmann constant and C is the constant.

The resins consist of a number of fractions with different molecular masses, different vapour pressures and different rates of evaporation, which makes it unpractical to apply the Langmuir equation. Though, this equation is suitable for providing some qualitative information on the temperature and pressure effects on the evaporation rate, a more quantitative expression for the reaction velocity of the polymerisation process includes the chemical activation, the polymerisation reaction velocity and the polymerised fraction. In its simplest form it can be expressed as [66,67]:

$$\partial\beta/\partial t = C_t(1-\beta)^n = Ae^{-E/kT}(1-\beta)^n \tag{7.3}$$

where β is the degree of polymerisation, $\partial\beta/\partial t$ is the reaction velocity (L/mol s); A is the constant; E is the chemical activation energy (kJ/mol); k is the Boltzmann constant and T is the temperature (K).

The modelling of the competition between the chemical reaction (self-healing of monomer with the catalyst) and the evaporation process could involve also additional parameter such as the concentration of the catalyst:

$$\partial C_1/\partial t = -div(D_1 gradC_1) - R \tag{7.4}$$

$$\partial C_2/\partial t = -div(D_2 gradC_2) - R \tag{7.5}$$

$$\partial\beta/\partial t = R \tag{7.6}$$

$$R = k(1-\beta)(1+\alpha\beta)(1-\gamma\beta) \tag{7.7}$$

where the indices 1 and 2 refer to the two components (e.g., resin and hardener, or monomer and catalyst); C_1 and C_2 refer to concentration; D_1 and D_2 refer to diffusion rate (cm^2/s); β refers to the degree of poly-merisation; $R = \partial\beta/\partial t$ refers to reaction velocity; α refers to catalyst or auto-catalyst factor and γ refers to auto-slowdown factor (includes the evaporation effect).

The effect of the space plasma (AO, VUV and radiation) may be ambivalent, that is, beneficial to curing (such as polymerisation and curing using solar UV light) or damaging and degrading the used materials (e.g., the breaking of polymer molecules by high energy particles). This

ambivalent effect of space radiation makes the polymer curing more difficult to predict.

The use of MM-374 resin was proposed for space inflatable structures
[66,67]. The evolution of its mass at different curing temperatures and at
different vacuum pressures was measured (Figure 7.15). The loss due to
outgassing is small (at 80 °C and 120 °C) unless it is used at the higher
temperature (160 °C). It is common to find some polymers losing 1%–5%
of their weight during the mission lifetime [66].

The chemical reaction rates were estimated and measured from the data
obtained from the IR absorption spectroscopy performed on the MM-374
resin and its associated hardener material (Figures 7.16 and 7.17).

The evolution of the IR lines associated to the resin and hardener as
a function of the time has found to follow an S-shape (Figure 7.17). The
S-shape of these curves corresponds to a second order polymerisation
reaction. However, it is possible to estimate roughly number of data related
to the reaction, including the auto-catalyst and slowdown parameters.

The evaporation effect can be indeed included in the slowdown parameter. The auto-catalyst and slowdown parameters, α and γ, respectively,
can be estimated from comparing the modelling data with the experimental
results for the inflexion point and the asymptotic values on the S-shaped
absorbance curve. Even if this method is widely used, in addition to require

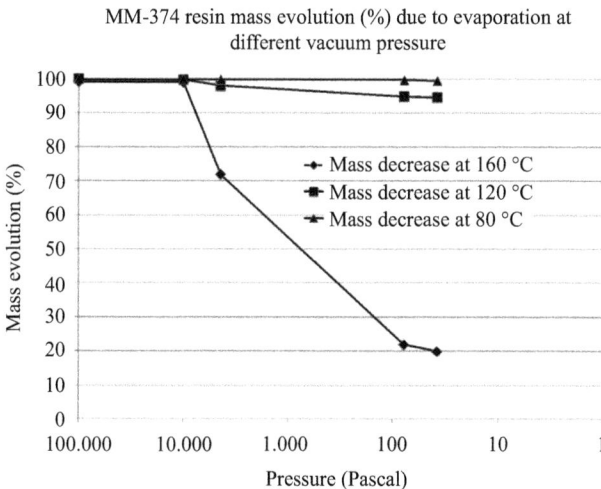

*Figure 7.15   Experimental results of the variation of the MM-374 resin
mass, after 1 hour, for different temperatures and vacuum
pressure levels. © 2003 European Space Agency
[Reproduced, with permission, from [66]]*

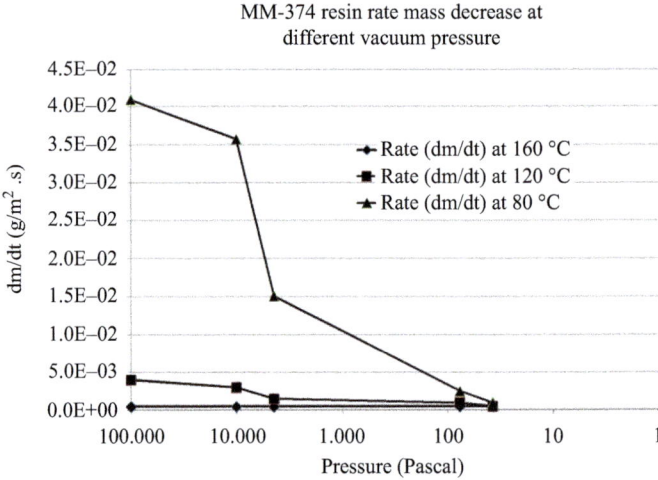

Figure 7.16   Experimental results of the variation of the polymerisation rate of the resin MM-374, for different temperatures, after 1 hour, at different vacuum pressure levels. © 2003 European Space Agency
[Reproduced, with permission, from [66]]

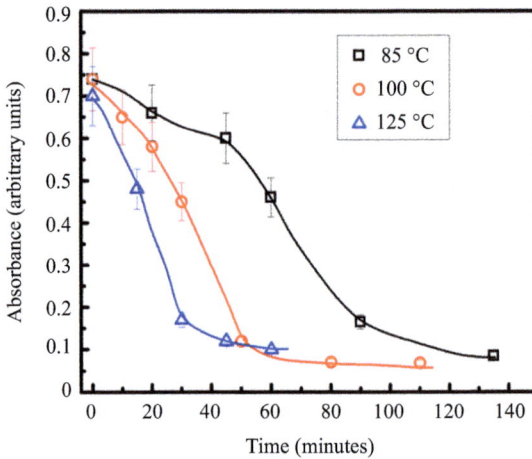

Figure 7.17   Experimental results of the variation of the polymerisation rate in time of the resin MM374 (measured at three different temperatures 85, 100 and 125 °C, by the absorptance of IR identified lines. ©2003 European Space Agency
[Reproduced, with permission, from [66]]

IR study under high vacuum, it cannot provide accurate values for α and γ but rather just an approximation.

The modelling of the self-healing process at low and high temperatures is more challenging, because the competition healing reaction (polymerisation) and evaporation becomes more complex:

- At low temperatures the evaporation of the monomer is slow, but at the same time the healing is also very slow (taking several hours) and
- At high temperatures, the evaporation of the monomer is high, but at the same time the healing is also very fast (taking only a few seconds).

The ENB monomer evaporates partially under vacuum (about 35% in free microcapsules), but less than 0.2% in a tapered double-cantilever beam standard sample, where the monomer may only leak from the microcapsules very close to the surface of the resin (Figure 7.18 and Table 7.9).

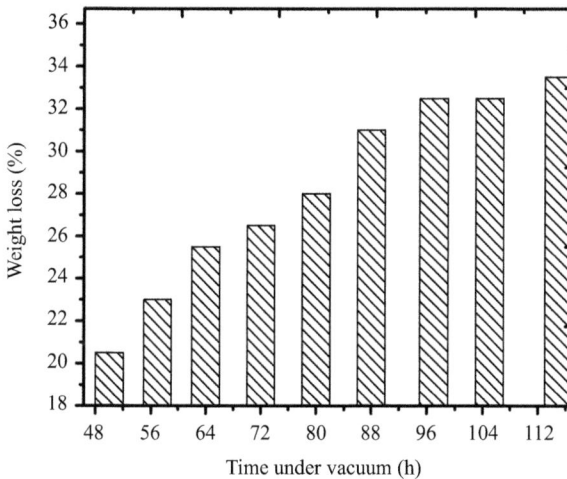

Figure 7.18   Weight loss due to the outgassing in vacuum of free microcapsules

Table 7.9   Loss of weight in vacuum (100 Pa) at room temperature (23 °C)

Experiment	Weight loss	Duration in vacuum
Free microcapsules not embedded	Approximately 35%	Almost stable after 5 days
Small device (3 g)	0.65%	Almost stable after 4 days
Standard device (33 g)	0.2%	Almost stable after 20 days (mainly due to microcapsules close to surface)

The self-healing process was tested in high vacuum conditions that are similar to a space environment (10^{-1}–10^{-2} Pa). Figure 7.19 illustrates the experimental set up. A sample holder was built with a platform allowing the heating and cooling of the self-healing demonstrator under vacuum. In order to increase the thermal conductivity and ensure homogeneous heating, the self-healing samples were embedded between two aluminium plates and two heating foils.

The results showed a stable healing efficiency (around 75%) for all the measurements under vacuum and in air at different temperatures between $-20\,^\circ$C and $+60\,^\circ$C.

(a)

(b)

(c)

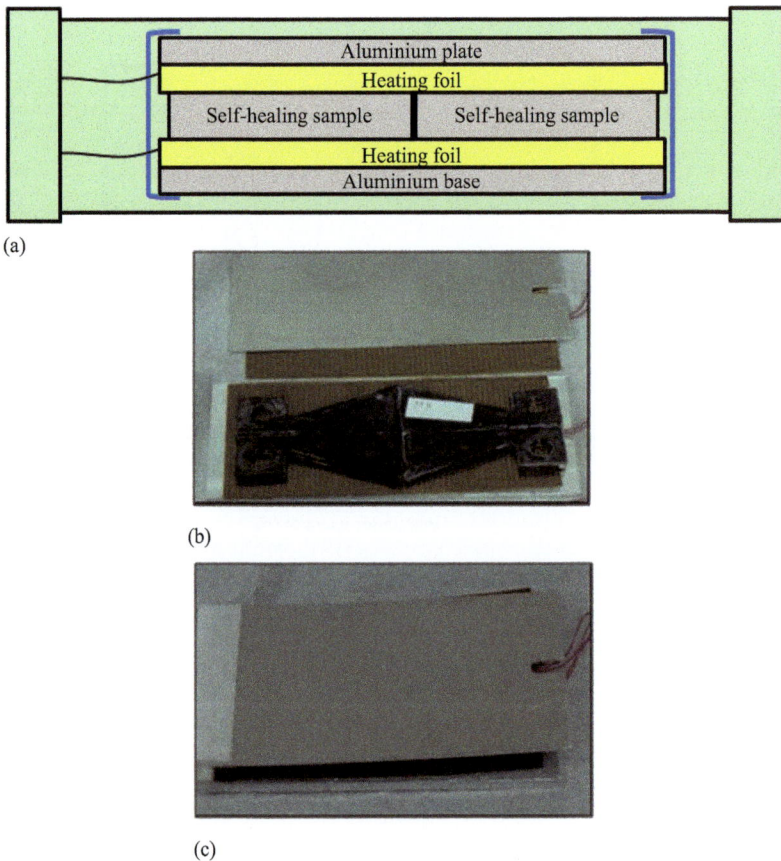

Figure 7.19   (a) Schematic and (b–c) photographs of the sample holder used for a self-healing test under vacuum

7.3.5 Space plasma

In addition to the AO, the space plasma includes UV radiation and atomic particles (electrons, protons, gamma radiation and high energy nuclear particles). The plasma may degrade the undergoing chemical process by breaking the polymer molecules. For example, the effect of protons can cause a local heating up to 9727 °C, and it can break the chemical bonds in a molecule and liberate carbon and hydrogen ions [66,67]. The high-energy particles may also break the microcapsule and cause the flow of the monomer.

7.3.6 Thermal shock

Thermal shock, induced by sudden and extreme changes in ambient temperature, may have additional destructive impact on structures used in space. Therefore, it is necessary to test any structure or device to assess the levels of stress that a material can withstand before a failure occurs. The test for thermal shocks has been applied to reveal mechanical weakness, delays in healing or noncompatibility between the ENB monomer, the catalyst and the host resin structure. The applied thermal shock has much more stringent conditions (−195 °C to 60 °C) than the standard thermal cycling tests. Liquid nitrogen and heating plates are used in this test. Commonly used thermal shock tests are based on the MIL-STD-202 Method 107 [69]. The high temperature is obtained by placing the sample within an oven kept at a fixed high temperature.

The thermal shock test conditions are summarised in Table 7.10.

Thermal shock was performed using either:

- Dual chamber, liquid-nitrogen (LN_2) to water (liquid-liquid) system or
- Dual chamber, LN_2/oven-plate (liquid-air).

Figure 7.20 illustrates one of the instruments used for the thermal shock test. The sample survived the 20 cycles of thermal shocks.

Table 7.10 Summary of the thermal shock test conditions

Parameter	MIL-STD-883 [70]	Test
Low temperature	−135 °C	−195 °C (liquid nitrogen)
High temperature	150 °C	60 °C (resin can melt for temperatures >80 °C)
Dwell time at each extreme	10 min	10 min
Time between low and high temperature	<1 min	<1 min
Number of cycles	10 cycles	20 cycles

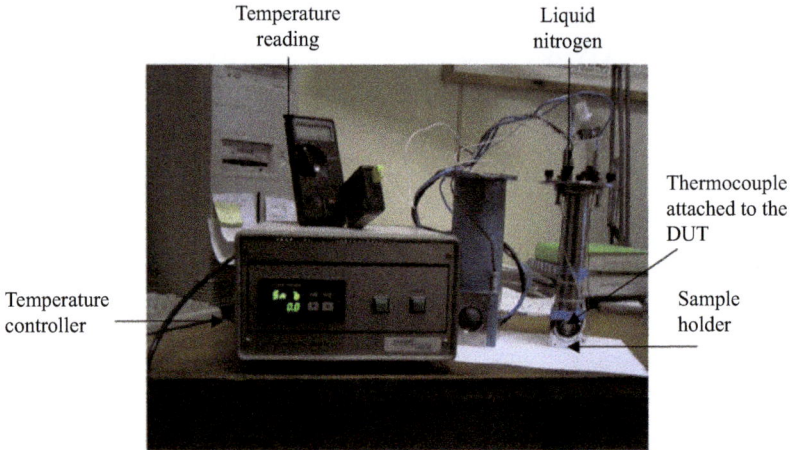

Figure 7.20   Photograph of the thermal shock setup. DUT, device under test

7.3.7   Outgassing

Outgassing (sometimes called offgassing) is the release of a gas that was dissolved, trapped, frozen or absorbed in the material [71]. Outgassing can include sublimation and evaporation, which are phase transitions of a substance into a gas. Boiling is generally thought of as a separate phenomenon from outgassing because it consists of a phase transition of a liquid into a vapour. Outgassing is a challenge to creating and maintaining clean high-vacuum environments. Outgassing products can condense onto optical elements, thermal radiators or solar cells and alter their functionality. NASA and ESA maintain a specific list of low-outgassing materials to be used for space applications. Materials not normally considered absorbent can release enough lightweight molecules to interfere with industrial or scientific vacuum processes. Moisture, sealants, lubricants and adhesives are the most common sources, but even metals and glasses can release gases from cracks or impurities. The rate of outgassing increases at higher temperatures because the vapour pressure and the rate of chemical reaction increases. For most solid materials, the method of manufacture and preparation can significantly reduce the level of outgassing. Cleaning surfaces, baking individual components or the entire assembly before use can drive off volatiles. In the high vacuum of space, composite materials, such as carbon/epoxy systems, can desorb moisture which can lead to large dimensional changes in the structure. The effect of moisture on composite laminates, for example, has been well documented in the literature [72–74]. Authors have noted that the dimensional changes caused by moisture desorption are highly influenced by the laminate design and the matrix material. Any selection of resin, composite and/or self-healing-based composite will have to take this parameter into consideration. It is worth noting here, that as well, the effect of moisture desorption on the composite's stability is often higher than that of the temperature variation. Consequently, the question of outgassing effects for self-healing composite materials has been considered in depth compared to the issue of the difference in the CTE. However, the extent of the outgassing effect that could occur in a structural composite is somehow a random phenomenon and depends on many factors such as the environment parameters that are not automatically taken into account during the initial design phase of the self-healing composite [75]. A comprehensive characterisation of outgassing effects using mass spectrometers has been compiled for ESA's Rosetta spacecraft [76].

References

[1]   ECSS-Q-70-04, *Space Product Assurance Thermal Cycling Test for the Screening of Space Materials*, 1999.

[2]   ECSS-Q-70-02, *Space Product Assurance, Thermal Vacuum Outgassing Test for the Screening of Space Materials*, 2000.

[3]   E.M. Silverman, *Space Environmental Effects on Spacecraft: LEO Materials Selection Guide*, NASA CP-4661 Part 1, NASA, Washington, DC, 1995.

[4]   *Self-Healing Materials: An Alternative Approach to 20 Centuries of Materials Science*, Ed., S. Van der Zwaag Springer Series in Materials Science, Volume 100, Springer, Dordrecht, the Netherlands, 2007.

[5]   S.D. Bergman and F. Wudl, *Journal of Materials Chemistry*, 2008, **18**, 1, 41.

[6]   R.P. Wool, *Soft Matter*, 2008, **4**, 3, 400.

[7]   D.Y. Wu, S. Meure and D. Solomon, *Progress in Polymer Science*, 2008, **33**, 5, 479.

[8]   M.R. Kessler, *Proceedings of the Institution of Mechanical Engineers, Part G: Journal of Aerospace Engineering*, 2007, **221**, 4, 479.

[9]   Y.C. Yuan, T. Yin, M.Z. Rong and M.Q. Zhang, *Express Polymer Letters*, 2008, **2**, 4, 238.

[10]   J.P. Youngblood, N.R. Sottos and C. Extrand, *MRS Bulletin*, 2008, **33**, 8, 732.

[11]   K.A. Williams, D.R. Dreyer and C.W. Bielawski, *MRS Bulletin*, 2008, **33**, 8, 759.

[12]   S.R. White, M.M. Caruso and J.S. Moore, *MRS Bulletin*, 2008, **33**, 8, 766.

[13]   I.P. Bond, R.S. Trask and H.R. Williams, *MRS Bulletin*, 2008, **33**, 8, 770.

[14]   R.S. Trask, H.R. Williams and I.P. Bond, *Bioinspiration and Biomimetics*, 2007, **2**, 1, 1.

[15]   S-K. Ghosh, Ed., *Self-healing Materials: Fundamentals, Design Strategies, and Applications*, Wiley-VCH Verlag, Weinheim, Germany, 2009.

[16]   C. Dry, *Composite Structures*, 1996, **35**, 3, 263.

[17]   S.R. White, N.R. Sottos, P.H. Geubelle, *et al.*, *Nature*, 2001, **409**, 6822, 794.

[18]   J.G. Smith, Jr., *An Assessment of Self-healing Fiber Reinforced Composites*, NASA/TM–2012-217325 NASA Langley Research Center, Hampton, VA, 2012.

[19]   W. Li, J.W. Buhrow, L.M. Calle, *Proceedings of the 3rd International Conference on Self-healing Materials*, NASA #20110008504_2011008866, Bath, UK, 2011.

[20]   T.K. O'Brien, *Assessment of Composite Delamination, Self-healing under Cyclic Loading*, NASA #20090028603_2009028341, Langley Research Center, Hampton, VA, 2009.

[21]   D. Eberly, R. Ou, A. Karcz and G. Skandan, *Self-healing Nanocomposites for Reusable Composite Cryotanks Applications for COPVs include Storage of Natural Gas and Liquid Hydrogen Fuel in Vehicles, and Marine Transport of Propane via Tanker Ships*, NASA # 20130013555, NEI Corporation for NASA Marshall Space Flight Center, NASA, Washington, DC, 2013.

[22]   B. Johnson, A. Caracci, L. Tate and D. Jackson, *Fabrication and Characterization of (MWCNT) and Ni-MWCNT) Repair Patches for CFRP*, NASA #20110016716_2011017757, NASA Kennedy Space Center, Cape Canaveral, FL, 2011.

[23]   S.V. Raj, M. Singh and R. Bhatt, *New Class of Self-healing Ceramic Composites (SHCCs), Composites for Aircraft Engine Applications, Applications include the Nuclear Power Generation Industry and Military Ships*, NASA# 20130009813_2013009221, John H. Glenn Research Center, Cleveland, OH, 2013.

[24]   J.A. Riedell and T.E. Easler, *Ceramic Adhesive and Methods for On-orbit Repair of Re-entry Vehicles*, NASA# 20130013565_2013013352, Johnson Space Center, Houston, TX, 2013.

[25]   J.A. Riedell and T.E. Easler, inventors; COI Ceramics Inc, assignee; US7628878B2, 2009.

[26]   K. O'Brien, M.W. Czabaj J.A. Hinkley, *et al.*, *Combining Through-Thickness Reinforcement and Self-healing for Improved Damage*

Tolerance and Durability of Composites, NASA/TM–2013-217988, Langley Research Center, Hampton, VA, 2013.

[27] C. Johnson and G. Spexarth, *Proceedings of the Annual Technical Symposium, Inflatable Structures: Test Results and Development Progress since TransHab*, Document ID 20060022083, NASA Johnson Space Centre, Houston, TX, 2006, p. 20.

[28] A. Haight, J-M. Gosau, A. Dixit and D. Gleeson, *Self-healing, Inflatable, Rigidizable Shelter - (Rigidization on Command)*, Marshall Space Flight Center, Huntsville, AL, 2012.

[29] E.J. Brandon, M. Vozoff, E.A. Kolawa, *et al.*, *Acta Astronautica*, 2011, **68**, 7-8, 883.

[30] W. Li and L.M. Calle, *Proceedings of the 210th Electrochemical Society Meeting*, Pennington, NJ, 2006, p. 143.

[31] L.M. Calle, J.W. Buhrow and S.T. Jolley, *SAMPE Fall Technical Conference*, Session 151-AB, Salt Lake City, UT, 2010.

[32] G.A. Slenski and P.S. Meltzer, Jr., *The AMPTIAC Quarterly*, 2004, **8**, 3, 17.

[33] S.T. Jolley, M.K. Williams, T.L. Gibson and A.J. Caraccio, *Next Generation Wiring Developing Flexible: High Performance Polymers with Self-healing Capabilities*, NASA Kennedy Space Center, FL, 2011.

[34] M.W. Keller, N.R. Sottos and S.R. White, inventors; University Of Illinois, assignee; US7569625 B2, 2009.

[35] D. R. Huston, and B. R. Tolmie, *Self-healing Cable for Extreme Environments*, US Patent, US20080283272 A1, 2008.

[36] J.I. Kleiman, Y. Gudimenko, Z.A. Iskanderova, R.C. Tennyson and W.D. Morison, *Proceedings of 4th International Space Conference on the Protection of Space Materials from the Space Environment*, University of Toronto, Toronto, Canada, 1998, Kluwer Academic Publishers, Dordrecht, the Netherlands, 2001, p. 243.

[37] A.M. Grande, A. Rahaman, L. Di Landro, M. Penco and I. Peroni, *Proceedings of the 4th European Conference for Aerospace Sciences EUCASS, organized by the European Scientists and Engineers*, Moskovskie Morota Hotel, St Petersburg, Russia, 2011.

[38] L. Castelnovo, A.M. Grande, L. Di Landro, G. Sala, C. Giacomuzzo and A. Francesconi, *Proceedings of the 4th International Conference on Self-healing Materials (ICSHM2013)*, Ghent, Belgium, 2013, p. 337.

[39] E.J. Siochi, J.B. Anders, Jr., D.E. Cox, D.C. Jegley, R.L. Fox and S.J. Katzberg, *Biomimetics for NASA Langley Research Center*, 2000 Report of Findings from a Six-Month Survey, NASA/TM-2002-211445, Langley Research Center, Hampton, VA, 2002.

[40]   T.A. Plaisted, A.V. Amirkhizi, D. Arbezlaez, S.C. Nemat-Nasser and S. Nemat-Nasser, *Smart Structures and Materials 2003: Industrial and Commercial Applications of Smart Structures Technologies*, Ed., E.H. Anderson, *SPIE Proceedings Volume 5054*, Bellingham, WA, 2003, p. 372.

[41]   L.A. Gavrilov and N.S. Gavrilova, *IEEE Spectrum*, 2004, **41**, 9, 30.

[42]   K.L. Bedingfield, R.D. Leach and M.B. Alexander, Eds., *Spacecraft System Failures and Anomolies Attributed to the National Space Environment*, NASA Reference Publication 1390, NASA, Marshall Space Flight Centre, Huntsville, AL, 1996.

[43]   G.A. Graham and A.T. Kearsley, *ESA-ESTEC - Space Environments and Effects - Final Presentation Days*, organised and held at ESA/ESTEC, Contract No. 13308/98/NL/MV, Noorddwijk, the Netherlands, 1999.

[44]   *ESA Space Debris Mitigation Handbook, IADC Protection Manual*, Documents No: IADC-02-01 and IADC-04-06. *http://www.iadc-online.org*

[45]   N.L. Johnson, *NASA Green Engineering Masters Forum*, San Francisco, CA, NASA, FL, 2009.

[46]   M.D. Rhodes, J.G. Williams and J.H. Starnes, Jr., *Proceedings of the 34th Annual Conference of the Reinforced Plastics/Composites Institute – Reinforcing the Future*, New Orleans, LA, 1979, p. 20D1.

[47]   J.D. Winkel and D.F. Adams, *Composites*, 1985, **16**, 4, 268.

[48]   C.K.L. Davies, S. Turner and K.H. Williamson, *Composites*, 1985, **16**, 4, 279.

[49]   J-K. Kim and M-L. Sham, *Composites Science and Technology*, 2000, **60**, 5, 745.

[50]   F. Mili and B. Necib, *Composite Structures*, 2001, **51**, 3, 237.

[51]   C.H. Yew and R.B. Kendrick, *International Journal of Impact Engineering*, 1987, **5**, 1-4, 729.

[52]   W.P. Schonberg, *Advances in Space Research*, 2009, **45**, 6, 709.

[53]   V.V. Silvestrov, A.V. Plastinin and N.N. Gorshkov, *International Journal of Impact Engineering*, 1995, **17**, 4-6, 751.

[54]   R.C. Tennyson and C. Lamontagne, *Composites Part A: Applied Science and Manufacturing*, 2000, **31**, 8, 785.

[55]   C.G. Lamontagne, G.N. Manuelpillai, J.H. Kerr, E.A. Taylor, R.C. Tennyson and M.J. Burchell, *International Journal of Impact Engineering*, 2001, **26**, 1–10, 381.

[56]   Y. Tanabe and M. Aoki, *International Journal of Impact Engineering*, 2003, **28**, 10, 1045.

[57]   M. Wicklein, S. Ryan, D.M. White and R.A. Clegg, *International Journal of Impact Engineering*, 2008, **35**, 12, 1861.

[58] A.R. Chambers, M.C. Mowlem and L. Dokos, *Composite Science and Technology*, 2007, **67**, 6, 1235.

[59] B. Banks, S.K. Miller and K.K. de Groh, *Proceedings of the 2nd AIAA International Energy Conversion Engineering Conference*, Providence, RI, 2004.

[60] R.V. Kruzelecky, A.K. Ghosh, E. Poiré and D. Nikanpour, *Protection of Materials and Structures from Space Environment (ICPMSE-4)*, Eds., J. Kleiman and R.C. Tennyson, Kluwer, Dordrecht, the Netherlands, 1998, p. 125.

[61] T.K. Minton, M.E. Wright, S.J. Tomczak, *et al.*, *ACS Applied Materials & Interfaces*, 2012, **4**, 2, 492.

[62] S.J. Tomczak, M.E. Wright, A.J. Guenthner, *et al.*, *Proceedings of the American Institute of Physics International Conference: Materials Physics and Applications*, Volume 1087, Toronto, Canada, 2008, p. 505.

[63] A. de Rooij, *Proceedings of the 3rd ESA European Symposium on Spacecraft Materials in Space Environment (ESA SP-232)*, Noordwijk, the Netherlands, 1985.

[64] Z. Iskanderova, J. Kleiman, Y. Gudimenko, R.C. Tennyson and W.D. Morison, *Surface and Coatings Technology*, 2000, **127**, 1, 18.

[65] K.K. de Groh, B.A. Banks, G.G. Mitchell, *et al.*, *Proceedings of the 12th ESA/ESTEC International Symposium on Materials in the Space Environment (ISMSE-12)*, Noordwijk, the Netherlands, 2012.

[66] A. Kondyurin and B. Lauke, *Proceedings of the European Space Agency 9th International Symposium on Materials in a Space Environment*, ESA SP-540, Noordwijk, the Netherlands, November 2003, p. 75.

[67] A. Kondyurin and B. Lauke, *Proceedings of the 2nd European Workshop on Inflatable Space Structures*, Tivoli, Italy, 2004.

[68] B.S. Bokshtein, M.I. Mendelev and D.J. Srolovitz, *Thermodynamics and Kinetics in Materials Science: A Short Course*, Oxford University Press, Oxford, 2005.

[69] MIL-STD-202G, *Department of Defense, Test Method Standard, Electronic and Electrical Component Parts, Method 107G Thermal Shock*, 2002. *http://snebulos.mit.edu/projects/reference/MIL-STD/MIL-STD-202G.pdf*

[70] MIL-STD-883E, *Department of Defense, Test Method Standard*, Microcircuits, 1996. *http://scipp.ucsc.edu/groups/fermi/electronics/mil-std-883.pdf*

[71] E. Miyazaki, M. Tagawa, K. Yokota, R. Yokota, Y. Kimoto and J. Ishizawa, *Acta Astronautica*, 2010, **66**, 5–6, 922.

[72] K. Chane-Ching, M. Lequan, R.M. Lequan and F. Kajzar, *Chemical Physics Letters*, 1995, **242**, 6, 598.

[73] P.K. Mallick, *Fiber Reinforced Composites: Materials, Manufacturing and Design*, 2nd Edition, Marcel Dekker, New York, NY, 1993.

[74] F.N. Cogswell, *Thermoplastic Aromatic Polymer Composites*, Butterworth-Heinemann Ltd, Oxford, 1992.

[75] C.O.A. Semprimoschnig, *Enabling Self-Healing Capabilities – A Small Step to Biomimetic Materials*, Materials Report No. 4476, ESA Technical Note, European Space Agency, Cologne, Germany, 2006. *http://esamultimedia.esa.int/docs/gsp/materials_report_4476.pdf*

[76] B. Schläppi, K. Altwegg, H. Balsiger, *et al.*, *Journal of Geophysical Research, Part A: Space Physics*, 2010, **115**, A12, A12313.

Chapter 8

Self-healing capability against impact tests simulating orbital space debris

The presence in space of micrometeoroids and orbital debris, particularly in the lower Earth orbit, presents a continuous hazard to orbiting satellites, spacecraft and the International Space Station. Space debris includes all nonfunctional man-made objects and fragments in Earth orbit. As the amount of debris continues to grow, the probability of collisions that could lead to potential damage will consequently increase. In this chapter, the feasibility of self-healing of impacted composites in space is discussed.

The use of carbon fibre reinforced polymers (CFRP) in space structures has expanded in the last few years. This can be seen by the number of papers dedicated to study their reliability, health monitoring in space and their response to debris. Typical satellite service modules are square or octagonal boxes with a central cone/cylinder and shear panels (SP). The cone cylinder and SP are usually constructed from a sandwich panel with CFRP face sheets and an aluminium (Al) honeycomb (HC) core (CFRP/Al-HC-SP). Similarly, the upper and lower platforms are also CFRP/Al-HC-SP. The lateral panels of the service module are, because of thermal reasons, sandwich panels with Al face sheets and Al-HC cores. These panels are also wrapped with multilayer insulation blankets.

Lateral panels may be made with CFRP/Al-HC-SP. Other payloads include telescopes, which are quite often constructed predominantly from CFRP (for stability and pointing requirements). Truss-type structures, used for supporting antennas, solar arrays and so on, are typically made of CFRP.

As described in the chapters 4–7, the self-healing process is based on microcapsules filled with various combinations of 5-ethylidene-2-norbornene (ENB) and dicyclopentadiene (DCPD) monomers, reacted with the ruthenium Grubbs' catalyst (RGC). The self-healing materials were then successfully mixed with an Epon® 828-based resin epoxy and single-walled carbon nanotube(s) (SWCNT) materials. They are infused into the layers of woven CFRP. Although the microcapsules would not heal the impact's crater zone, the healing of delamination developed around the crater over distances much larger than the crater diameter itself is feasible. The CFRP specimen structures were subjected to hypervelocity impact conditions, prevailing in the space environment, using an advanced implosion-driven hypervelocity launcher. The impacted CFRP specimens were systematically characterised using three-point bending tests for flexural strength evaluation. The self-healing efficiency and the SWCNT contribution were determined [1].

8.1   Elaboration of self-healing in resin and carbon fibre reinforced plastics

Impact of space debris was studied experimentally using hypervelocity pellets with a diameter between 1 mm and 5 mm, and velocity between 2 km/s and 8 km/s. Self-healing of cracks in the resin sample subjected to a high-velocity impact of metal projectiles was demonstrated by visual observation of the cracks. Self-healing resin samples were prepared by mixing the healing ingredients (microcapsules and Grubbs' catalyst) into the epoxy resin cured at room temperature. The resin sample was then exposed to a metal projectile from a high pressure gas gun. The damaged region on the resin sample was then analysed using an optical microscope.

8.1.1   Preparation of resin sample

Self-healing samples made of resin (Epon® 828), curing agent (Epikure™ 3046), microencapsulated ENB and Grubbs' catalyst were prepared according to the process shown in Figure 8.1. Figure 8.2(a) shows an optical image of the typical self-healing sample based on Epon 828 resin, and self-healing materials based on ENB/poly-melamine urea-formaldehyde (PMUF) microcapsules and Grubb's catalyst. One can note the well dispersion of the microcapsule inside the epoxy material (Figure 8.2(b)). The dimensions of the sample were $50.8 \times 50.8 \times 4.8$ mm.

8.1.2   Validation of the high velocity impact test on epoxy based samples

The resin samples shot by a metal projectile [2,3] are shown in Figure 8.2. The projectile is launched through the muzzle by compressed gas until the rupture of a diaphragm (see Figure 8.3). The compressed gas from the connected cylinder is accumulated behind the diaphragm, which under certain pressure, bursts and allows the projectile to travel through the launch tube. The speed of the projectile is controlled by varying the experimental parameters, such as gas pressure, thickness of the diaphragm and so on.

The specifications for the high velocity impact test conducted on the resin samples to create the intended damage were projectile: Al; diameter: 1.8 mm; mass: 15 mg and speed: 600 ± 50 m/s. The impact produces a

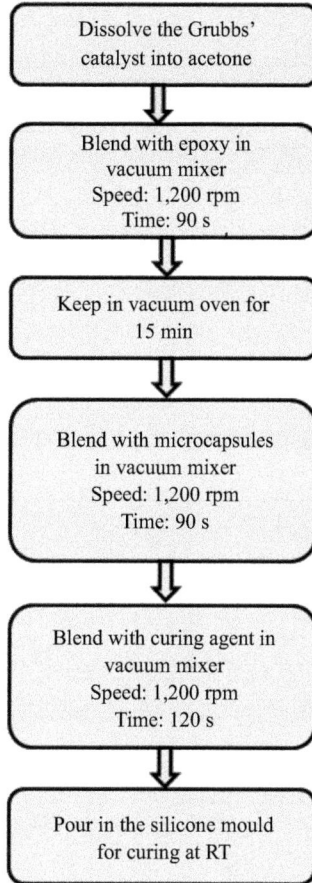

Figure 8.1   Processing route to making self-healing resin sample.
RT: room temperature

network of cracks in the sample as shown in the optical micrographs of Figure 8.4. Initial tests were carried out with epoxy samples. Afterwards, the test was performed with CFRP samples.

Some of the locations of the sample as observed under the microscope at different times are shown in Figure 8.5. These pictures should be contrasted with the appearance of the cracks.

It has been found that, at some locations, the cracks formed were completely filled with the polymer layer, indicating the self-healing of the cracks. No further changes in appearance of the cracks were observed as time elapsed. It appears that these cracks are healed (polymerisation of the healing agent released from the broken microcapsules into the cracks, the

(a)

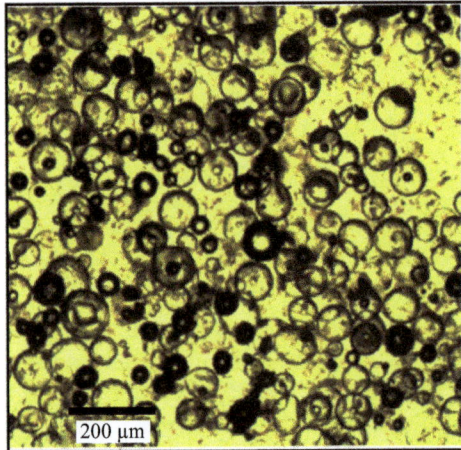

200 µm

(b)

Figure 8.2   (a) Optical micrograph of the typical self-healing sample made
of resin, curing agent and microcapsules and Grubbs' catalyst
and (b) dispersion of healing ingredients into the epoxy resin

release of the liquid healing agent (ENB) and its reaction with RGC)
rapidly, within just a few minutes, after the impact.

8.1.3   *Self-healing in carbon fibre reinforced polymer*
samples under high velocity impact

The data derived from the experiments with resin samples were then applied
to the next level of investigations, where the healing agents were

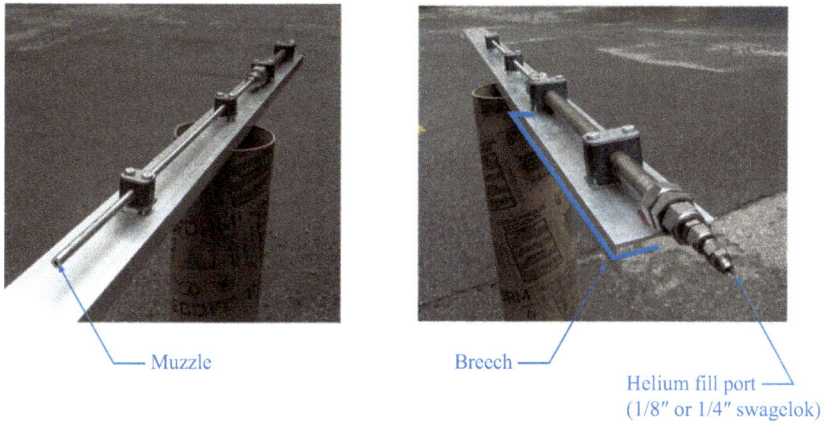

Muzzle   Breech

Helium fill port
(1/8″ or 1/4″ swagelok)

Figure 8.3 Single stage high velocity launcher

incorporated into the fibre-reinforced composites and tested with a high velocity impact. Several batches of separated individual microcapsules with an average size of less than 20 μm were synthesised using the method described earlier. Regular (without self-healing agents) and modified (with self-healing agents) cross ply $[0_2/90_2]_S$ composite panels were manufactured using the hand lay-up method with autoclave moulding. The microcapsules and Grubbs' catalyst were blended with Epon® 828 resin and then infused into the unidirectional layers of the carbon fibres to fabricate the self-healing composite panels. The same mixing scheme as that used with the resin sample was followed to infuse the self-healing agents into the modified CFRP samples.

A number of specimens with specified dimensions were cut off from each panel for further processing. Some of the specimens were exposed to high velocity projectiles. The specimens may be divided into four categories, that is, regular with no impact, regular exposed to the impact, self-healing with no impact and self-healed specimens exposed to the impacts. The flexural properties of the regular and modified specimens were then tested according to ASTM D7264 [4]. The self-healing performance was then evaluated by comparing the flexural strengths of the specimens. According to the dimensions of the specimen under test and based on the ASTM D7264 standard, a span to thickness ratio of 32 has been adapted. This ratio provides a plenty room for shooting, a large space for loading and support noses and a high degree of freedom for the overhang length.

Thus, this ratio is chosen for the three point bending test of the specimens. The impact conditions (projectile material, diameter, diaphragm thickness

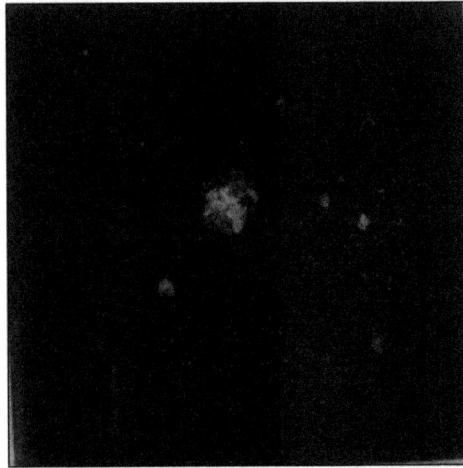

(a)

(b)

Figure 8.4   Optical photograph of the resin sample after impact. (a) Side of impact and (b) opposite side of the impact

controlling the burst pressure, velocity and so on) were appropriately chosen so that a given amount of created-damage, an intended level of damage, is guaranteed in the specimens. The impact conditions selected were projectile: Al; diameter: 4.27 mm and projectile speed: 450 ± 50 m/s. After shooting, the modified specimens were kept for 48 hours allowing time for self-healing.

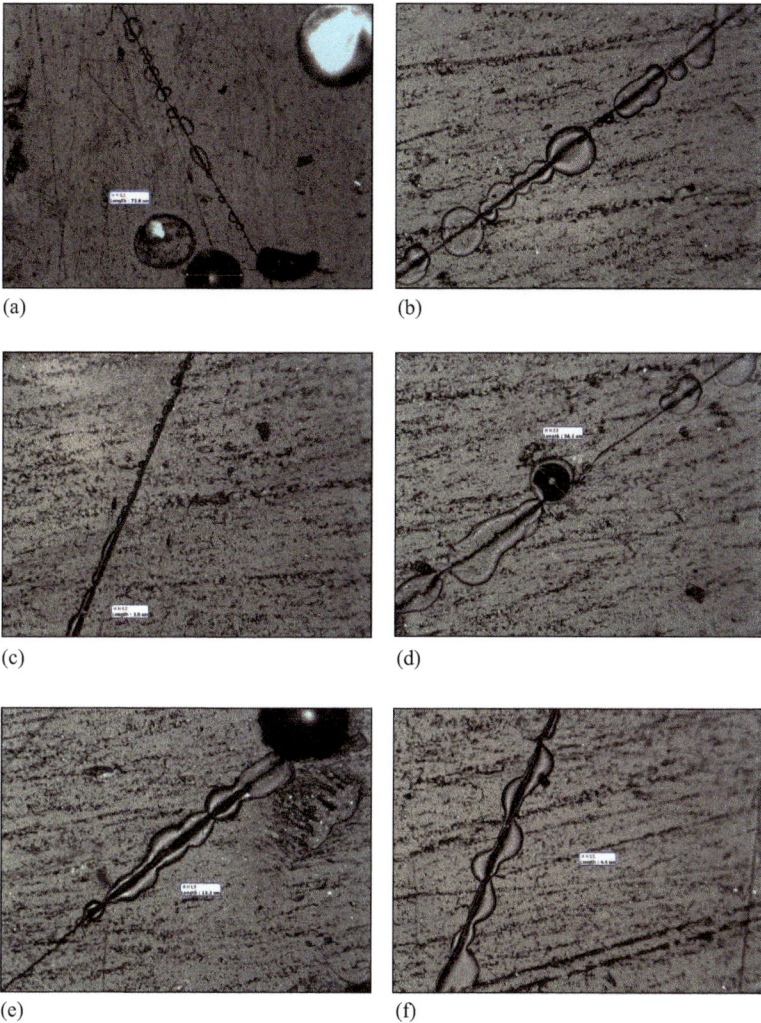

(a)   (b)

(c)   (d)

(e)   (f)

Figure 8.5   *(a–f) Self-healing of the samples as observed under an optical microscope at different locations. Observations have begun 15 minutes after the shooting test event*

The specimens were then tested for their flexural strengths using the three point bending specifications (Figure 8.6).

The measured flexural strengths were compared with those for regular (without self-healing) and modified (containing self-healing materials) specimens after impact and showed a self-healing recovery up to 54%.

Figure 8.6   Three-point bending test in progress

8.2   Self-healing in carbon fibre reinforced polymer samples under hypervelocity impact test

The same SWCNT as those in Chapter 9 were used and the same set of nanoscale characterisations were used in the tests. ENB and DCPD encapsulations were prepared following the same flowchart as described earlier. The self-healing demonstrator consisted of woven CFRP samples, containing three main constituents:

- Host matrix: an epoxy prepolymer (Epon® 828) and a curing agent (Epikure™ 3046) – this epoxy is used in space for internal structures;
- Microcapsules: the monomer healing agents (ENB/DCPD) were prepared as small microcapsules (diameter less than 15 μm). The monomer is homogeneously spread within the epoxy and forms about 10% of the structural weight;
- Different concentrations of SWCNT materials;
- Grubbs' catalyst first generation (ruthenium metal catalyst), 1 or 2 wt% and
- Ring-opening metathesis polymerisation (ROMP), that is, the healing chemical process that permits the reaction between the monomer and the catalyst.

8.2.1   Sample preparation

Different series of samples specimens were prepared, with and without carbon nanotube(s) (CNT). After the hypervelocity impact tests, the crack formed on the CFRP samples reached a microcapsule and caused its wall-rupture, which released the healing agent monomer (ENB or DCPD or a combination of the two monomers as will be described below) in the crack. Once the monomer and catalyst contacted, the self-healing reaction was triggered (i.e., polymerisation between the healing agent (monomer) and the matrix-embedded catalyst particles) [5–15]. The Grubbs' catalyst sustains the chemical reaction known as ROMP, up to the time the crack and the open microcapsules are full.

It is worth noting that the drawback to using the ENB monomer is that the resulting polymer is linear and thus, has inferior mechanical properties compared to DCPD. It was then proposed that the combination of these two monomers would give a fast, autonomic, self-repairing composite with excellent mechanical properties. Second, it was proposed to combine the known higher mechanical properties of CNT materials with these monomers to evaluate the self-healing capability. The microcapsules containing the healing agent monomer were embedded into woven CFRP (four woven layers were used, and the average size of the microcapsules was equal or less than 15 μm diameter). The fabrication procedure is summarised in Figure 8.7. The fibre Bragg grating (FBG) sensors were embedded between the second and third CFRP layers, concentrated inside a circle of 5 cm diameter (Figure 8.8) corresponding to the area exposed to debris during the hypervelocity impact experiment.

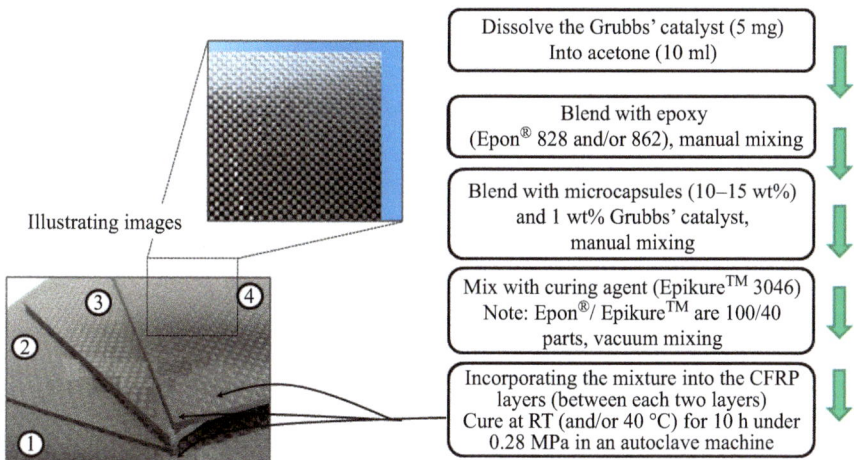

Figure 8.7   Fabrication procedure of manufacturing of CFRP with self-healing and FBG sensors

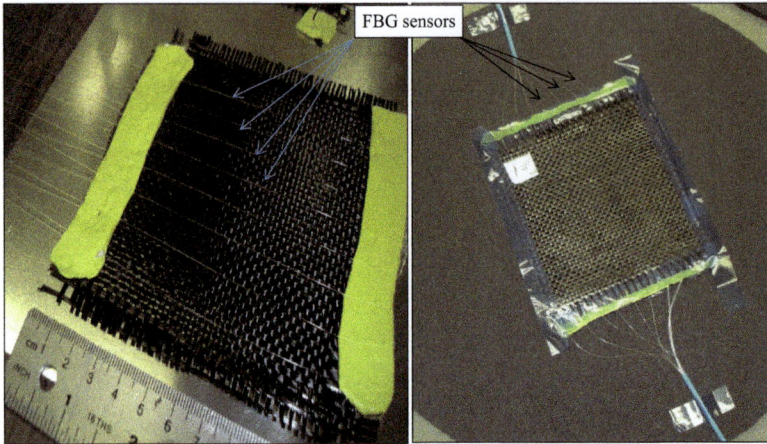

Figure 8.8   (a) Integration of four FBG sensors embedded between second
and third CFRP layers and concentrated inside a circle surface
of 5 cm diameter corresponding to the area exposed to debris
during the hypervelocity impact experiment, and (b) final
prototype of the sample

8.2.2   Hypervelocity impact test

Impact tests were performed with the implosion-driven hypervelocity launcher using the CFRP composite samples described previously. The sample (i.e., CFRP + FBG) was mounted in the target chamber, which was flushed with helium gas. The implosion-driven hypervelocity launcher functioned properly and to simulate the orbital space debris, small bore and/or polycarbonate projectiles (3–4 mm diameter) were launched with velocities reaching up to 8 km/s (Figure 8.9). All the CFRP samples fabricated were tested under the same conditions for comparison purposes. The impact resulted in complete penetration of the sample and a significant amount of delamination on both sides of the sample (Figures 8.12 and 8.13). In addition, frequent secondary impacts also occurred, which were likely to be the result of fragments generated by the hypervelocity launcher. Impacted CFRP samples were measured using the flexural three points bending test (ASTM D2344 [16]) after a healing process at 40 °C for 48 hours.

Seven sets of CFRP samples specimens (12 × 12 cm) were tested. The test data are summarised in Table 8.1. Figure 8.10 shows a CFRP sample after impact.

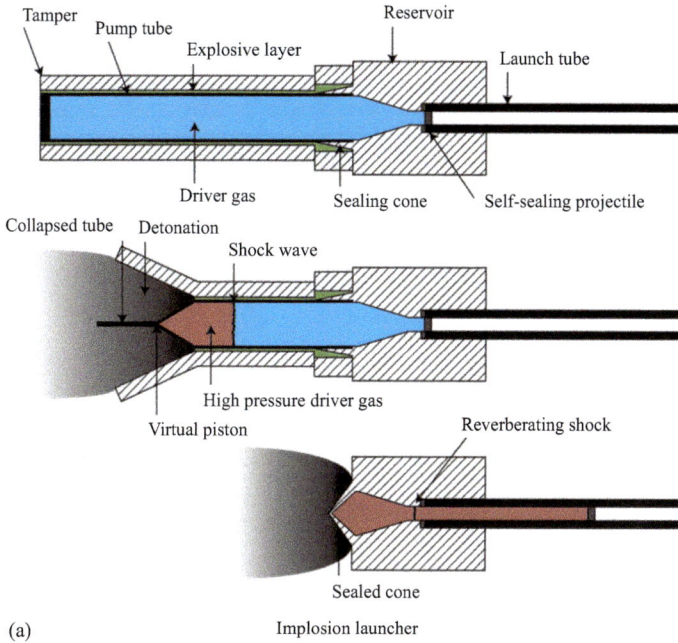

(a)   Implosion launcher

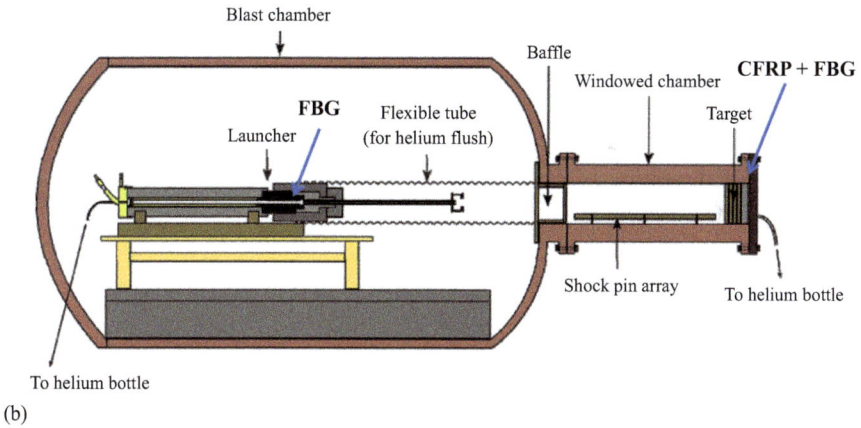

(b)

*Figure 8.9   (a) Review of implosion launcher concept and (b) schematic of
the launcher and the FBG experimental setup*

8.2.3   Study of the thickness of carbon fibre reinforced polymer samples after hypervelocity impact

Figure 8.11 shows a schematic of the CFRP sample after the hypervelocity impact test. The coordinates of the crater zone where the impact occurred

Table 8.1   Summary of the impacted CFRP structures (the self-healing
materials are mixed with 1 wt% of ruthenium Grubbs' catalyst
and 10 wt% of microcapsules)

Embedded material		Mechanical strength (MPa)	Healing efficiency (%)
Set # 1	Virgin Epon® 828	245	45
Set # 2	Epon® 828 + microcapsules of ENB	276	58
Set # 3	Epon® 828 + microcapsules of DCPD	290	63
Set # 4	Epon® 828 + microcapsules of (ENB + DCPD)	281	61
Set # 5	Epon® 828 + microcapsules of (ENB + DCPD) + 0.5 wt% CNT	297	68
Set # 6	Epon® 828 + microcapsules of (ENB + DCPD) + 1 wt% CNT	310	74
Set # 7	Epon® 828 + microcapsules of (ENB + DCPD) + 2 wt% CNT	326	83

Impact zone

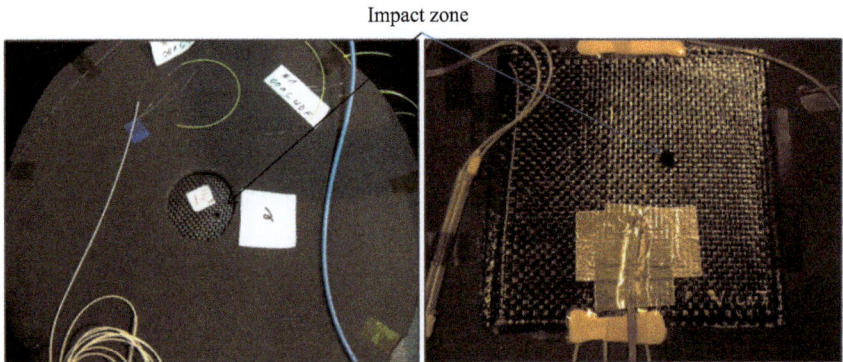

Figure 8.10   Typical example of the CFRP samples containing FBG
sensors after shooting under the hypervelocity impact

are first determined, then the sample is cut in slices having different widths, namely, 5 and 10 mm, depending on the distance to the crater zone (closer region are cut into smallest sizes of 5 mm). The goal is to investigate the information of the thicknesses of the slices as a function of their position to reproduce the mapping of the sample thickness after the hypervelocity impact event.

Figures 8.12–8.14 show the details of the thickness measurements. Sometimes metallic projectiles broke before the impact test and created two craters in the CFRP samples (Figure 8.14). Since the impact is supposed to produce delamination within the CFRP sample, we quantified these delaminations through their nominal thickness, which represents the ratio of the thickness

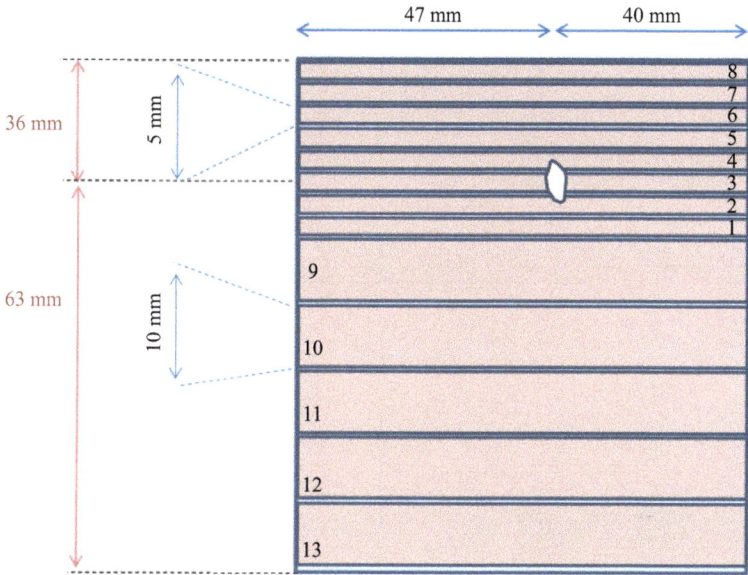

Figure 8.11   Schematic showing how the CFRP-FBG self-healing system was cut in slices after the impact event

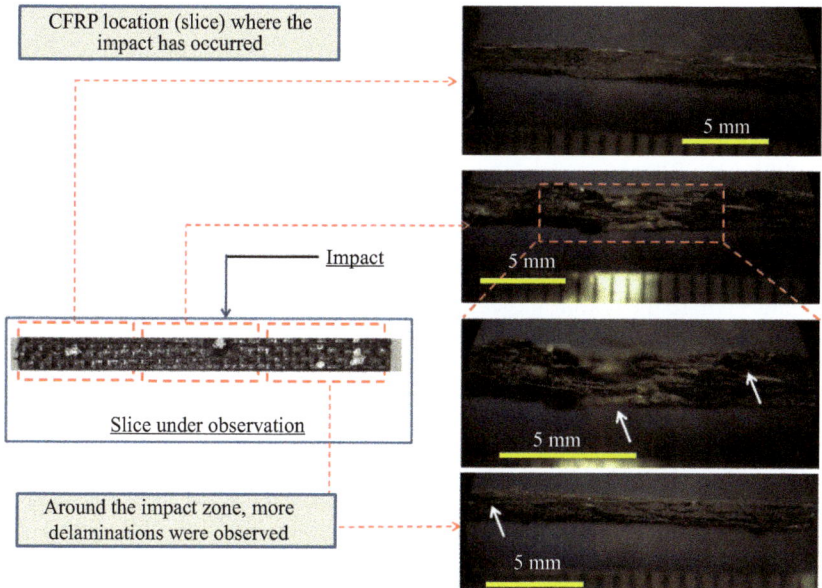

Figure 8.12   Example of the slice studies under the microscope

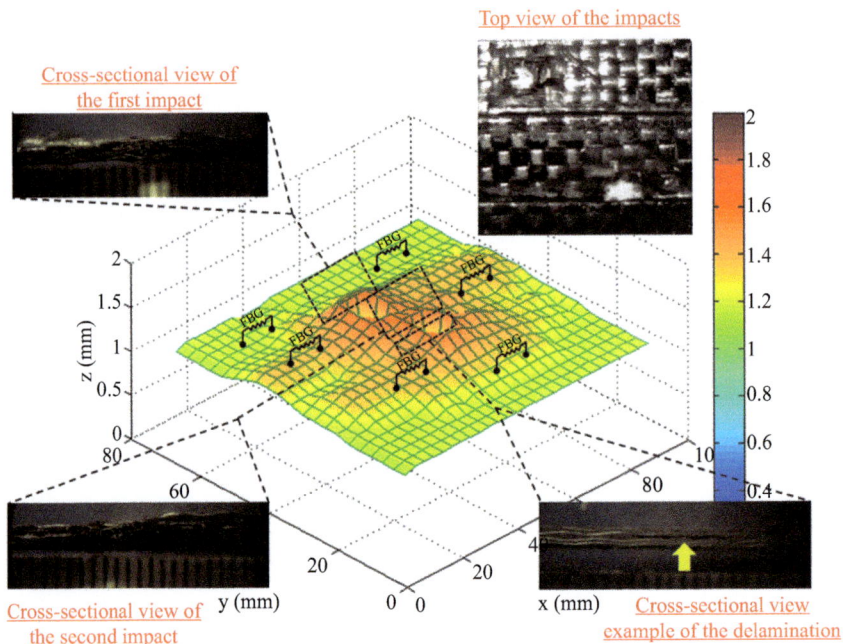

Figure 8.13   *Three-dimensional representation of the measured thickness of CFRP sample with two projectiles impact*

after/before impact. The thicker zone should then indicate where the delamination occurred. As mentioned before, a CFRP sample was cut into slices, and the matrix-thicknesses were estimated built by the measure of the thickness through the cross-sectional photos (the thickness is estimated for each 3 mm step in the two dimensions of the square sample; a matrix of more than 650 values is obtained for each case and each sample).

MATLAB® code was applied to trace two dimensions by using colour graphics where the colour indicates the width (thickness) of the zone (the wider zone is where the delamination is larger). Figure 8.14 shows a typical example of a colour code diagram obtained for CFRP-based Epon® 828, with and without self-healing materials.

There were two main observations:

- The colour code diagram indicates that the thicker portion of the sample containing delaminations is more localised around the impact zone, especially for the self-healing samples.
- The thickness distribution of the CFRP samples containing self-healing material seems to be much more homogenised compared to the distribution of the virgin one (i.e., pure Epon® 828), which indicates a clear healing effect (i.e., less delamination propagation after the healing process).

CFRP with pure resin

(a)

CFRP with self-healing material

(b)

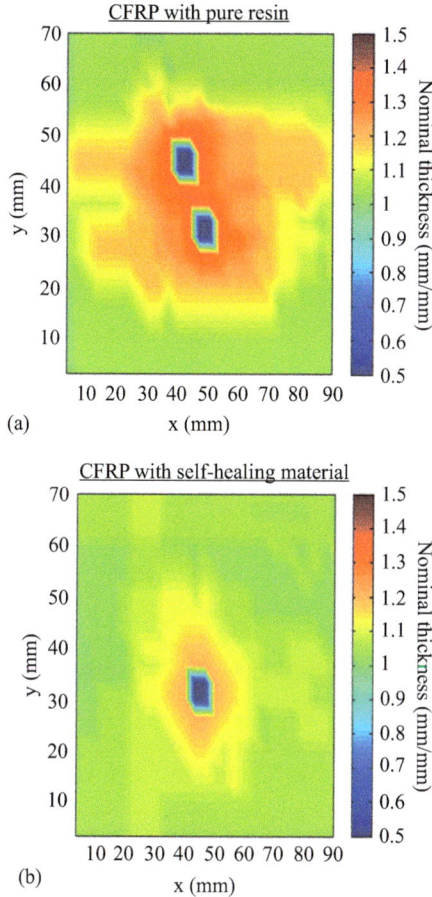

Figure 8.14   Typical example of a colour code diagram simulation of the impacted sample in cases (a) without and (b) with self-healing materials, showing that the thicker portion of the sample is mainly localised around the impact zone

8.2.4   Three point bending test

Impacted CFRP samples have been measured under the material testing system machine (three point bending test) after the healing process (48 hours at 40 °C) to investigate their mechanical properties. In accordance with ASTM D2344 [16], each sample was cut into 7–8 slices, as shown Figures 8.12 and 8.13.

Flexural three point bending measurements were performed twice on the CFRP samples (which were cut into slices, as described earlier):

- One measurement was made after the hypervelocity impact and healing process (48 hours, 40 °C) and
- The second measurement was performed on other bent slices (which were pressed horizontally under a press) after a second healing process

(48 hours, 40 °C). The mechanical flexural measurement (i.e., the three point bending test) was taken into the vertical line to the impact position. The data from the virgin CFRP samples (i.e., containing pure epoxy Epon® 828) before hypervelocity impact tests were also included for comparison purposes.

Three kinds of self-healing materials were used:

- One based on an encapsulated ENB monomer,
- The second based on the encapsulated DCPD monomer and
- The third was based on a 50/50 wt% mixture of these two monomers.

All the self-healing materials were blended with 1 wt% of RGC and 10 wt% of microcapsules.

The third class of the self-healing material was then mixed with different concentrations of CNT. The epoxy matrix was kept the same for all samples (namely, Epon® 828 with the Epikure™ 3046).

Table 8.1 summarises the mechanical strengths obtained from the flexural tests performed on the CFRP samples containing different healing agents and various SWCNT concentrations. The main goal was to determine the mechanical contribution of the healing part and that due to the CNT materials.

When comparing the test samples to the pristine samples (i.e., those containing only epoxy material), the recovery rate due exclusively to the healing materials was determined. The results may be summarised by the following conclusions:

- A mechanical strength of 31 MPa are due exclusively to the self-healing material based on ENB, which represents an enhancement of about 13%;
- When using the DCPD-based healing agent, better healing was obtained (the improvement was up to 18% of the mechanical strength);
- When using a mixture of 50/50 wt% of DCPD/ENB healing agent, a slight decrease of the flexural strength occurred (from 18% to ~15%), which was because of the incorporation of the ENB (ENB is a linear polymer having a lower mechanical strength, its addition to the DCPD slightly decreases the overall mechanical strength of the mixture);
- A clear improvement is obtained when integrating the SWCNT material, even with concentrations as low as 0.5 wt% and
- An enhancement up to 81 MPa for the mechanical recovery is found with the healing materials containing 2 wt% of SWCNT, which represents an improvement in the mechanical strength of about 33%.

When comparing between the healing efficiencies (estimated as a ratio of the flexural strength obtained from two sequential bending tests after the healing process), one can see that the healing efficiency because of the incorporation of only 2 wt% of SWNT can reach the value of 83%.

It is worth noting that the CNTs present in the crater zone area are supposed to disappear with the *ejecta*. However, those present in the delamination zones play a crucial role in the reconstruction of the material and help the self-healed composite to repair by acting as crosslinks in the released ENB polymer after its ROMP polymerisation with the Grubbs' catalyst.

8.2.5   Damping effects of the carbon nanotubes material

Figure 8.15 shows the response of the measured FBG centre wavelength (CWL) signals as a function of the time. These data were obtained for the impacted CFRP structures having a 50/50 wt% of ENB/DCPD healing agent and different concentrations of CNT. The inset of Figure 8.15 shows the extraction of FBG-CWL position as a function of the CNT loads. The samples with 1% and 2% SWCNT showed high damping effects due to the presence of a higher load of CNT. Work is under progress to correlate the mechanical contribution of the SWCNT materials to their morphological and structural properties.

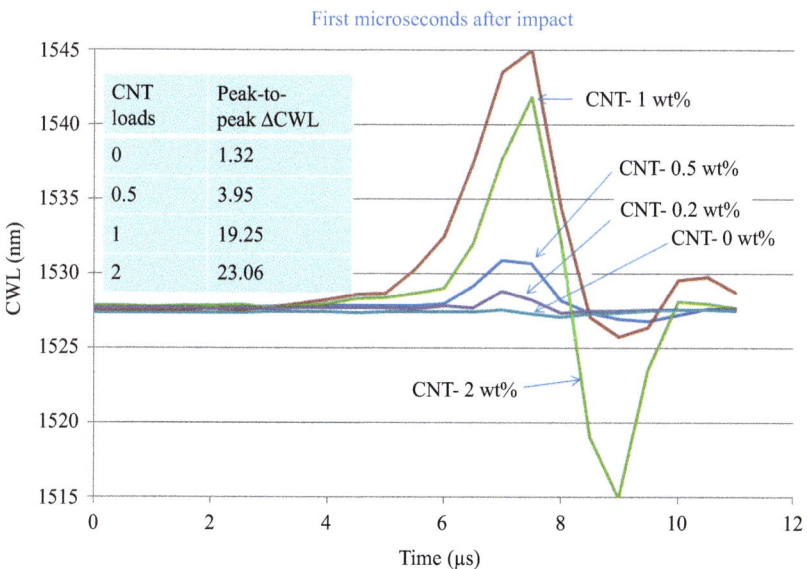

CNT loads	Peak-to-peak ΔCWL
0	1.32
0.5	3.95
1	19.25
2	23.06

Figure 8.15   FBG-CWL signals with respect to the time of the impacted CFRP structures having different CNT loads

8.3   Hypervelocity measurement with fibre Bragg grating sensors

Various CFRP embedded CNT and self-healing composite (based on microencapsulated ENB monomer and RGC embedded within the Epon® 828 epoxy system) samples were prepared. The FBG were attached on two locations, namely, the implosion tube and the output flange as follows:

- The FBG were attached on the implosion tube to measure:
  - The strain during the projectile launch and
  - Measurement of the projectile speed through the delay between two FBG responses (however, the FBG were too close to each other, i.e., only 5–10 cm, and the time delay was in the range of a few microseconds, which was too short for an accurate measurement).
- FBG sensors embedded within the CFRP and placed at the end of the chamber (Figure 8.16) and in this case the FBG measured:
  - The shock propagation within the sample,
  - The self-healing efficiency and
  - The projectile speed by measuring the delay between the FBG response on the tube and those on the CFRP samples.

Three FBG sensors were placed on the implosion tube and one FBG sensor on the CFRP sample. All sensors were interrogated with the high speed 2 MHz system.

The objective was to:

- Measure the pellet velocity, from the fibre sensor signal, and by comparing it to that measured by the streak camera. The camera measured a velocity of 7.9 km/s and
- Measure the strain evolution on the launching tube and to then compare it with the shock wave measurement.

Figure 8.17 shows the implosion tubes after the launch.

Figures 8.18 and 8.19 show the representative FBG signals obtained at the impact time.

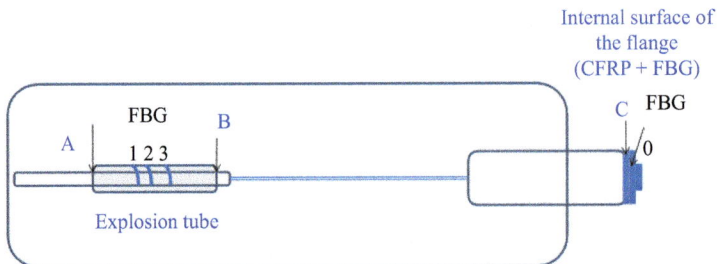

Figure 8.16   CFRP sample containing three FBG sensors on the launching tube, one FBG sensor within the CFRP

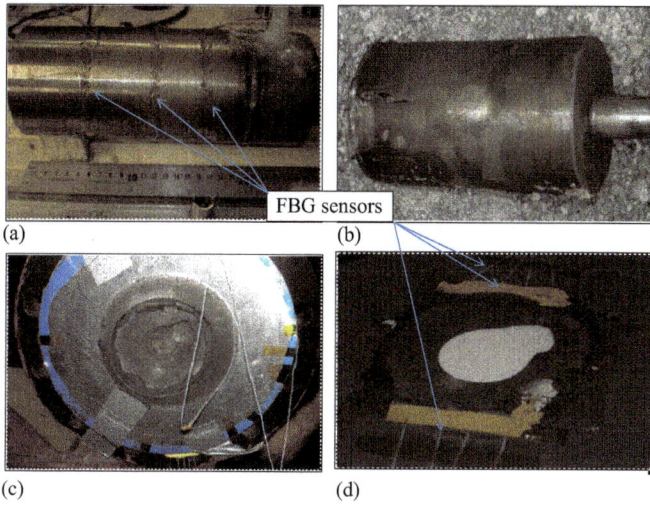

Figure 8.17 Optical photos of the launching implosion tube (a) before (with the FBG fibres glued); (b) after the impact test and (c–d) CFRP-FBG sample after the impact test

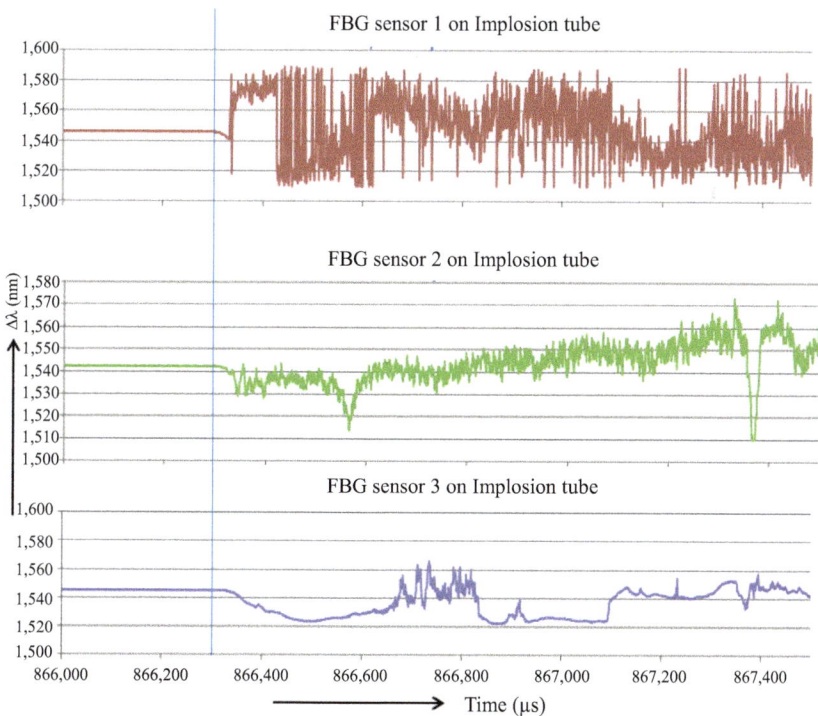

Figure 8.18 Typical recorded FBG signals obtained on the launcher implosion tube – the fibres broke after about 1 ms

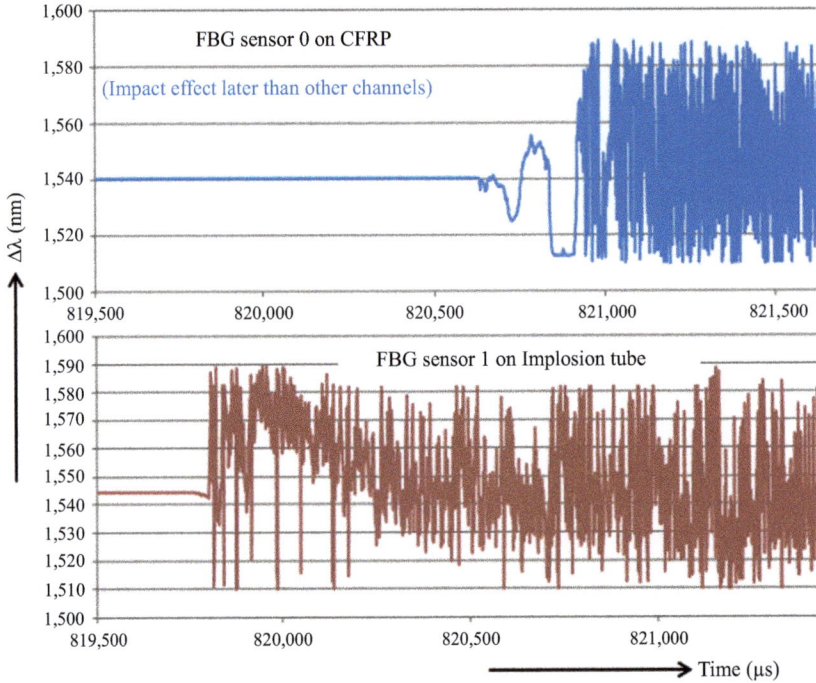

*Figure 8.19   Representative FBG responses on the implosion tube and
              CFRP sample, which occur about a half ms after the
              implosion event*

*Table 8.2   Pellet speed based on times of first change in the wavelengths.
            Average velocity 7.55 km/s (less than 4% error)*

Measured parameters	FBG0-FBG1	FBG1-FBG2	FBG2-FBG3	FBG1-FBG3
Delta-time (s)	0.0004175	0.0000025	0.000011	0.00000675
Delta-distance (m)	3.17	0.065	0.05	0.115
Local speed (m/s)	7,592.81	26,000.00	4,545.45	17,037.04
Speed (m/s)	7,592.81	7,481.93	7,675.28	7,437.61

The pellet velocity was calculated from the time difference recorded by
the FBG. The velocities were calculated based on the time delay of the
change of the wavelengths (Tables 8.2 and 8.3). The goal was to have a
good set of statistical data for the velocity estimation. In summary, the
embedded FBG sensors have shown a good capability for estimating both
the impact event and the pellet hypervelocity, with an accuracy around 1%.

Table 8.3   Pellet speed based on times of first jump. Average velocity 7.907 km/s (within 1% error)

Measured parameters	FBG0-FBG1	FBG1-FBG2	FBG2-FBG3	FBG1-FBG3
Delta-time (s)	0.000404	0.000009	0.0000185	0.00001375
Delta-distance (m)	3.17	0.065	0.05	0.115
Local speed (m/s)	7,846.53	7,222.22	2,702.70	8,363.64
Speed (m/s)	7,846.53	7,860.76	8,093.39	7,828.32

8.4   Summary of the hypervelocity impact study

It was demonstrated that self-healing materials may be used in a space environment. The auto-repair composite was a blend of microcapsules containing various combinations of ENB and DCPD monomers, reacted with RGC. Both monomers were encapsulated into a PMUF shell material, where the average size of the capsules was less than 15 μm. The self-healing materials were mixed with a resin epoxy, Epon® 828 and SWCNT by means of a vacuum centrifuging technique. The nanocomposites obtained were infused into the layers of woven CFRP. The CFRP specimens were then subjected to hypervelocity impact conditions using an advanced implosion-driven hypervelocity launcher. After the hypervelocity event, the three point bending tests show that the optimum self-healing material with the best mechanical strength was an equal-weight blend of ENB and DCPD monomers, while a huge improvement in terms of the auto-repair efficiency was obtained when adding small quantities of SWCNT (2 wt% and less). These results establish that using the ENB/DCPD/SWCNT/RGC system is a realistic possibility, which possesses tremendous potential for space applications. However, additional experimental investigations, especially a systematic cryomicrotome analysis as a function of the CNT load are needed to demonstrate the damping effect of the SCWNT material and their role as a crosslinks in the polymer formed.

References

[1]   B. Aïssa, K. Tagziria, E. Haddad, *et al.*, 'The self-healing capability of carbon fibre composite structures subjected to hypervelocity impacts simulating orbital space debris', *ISRN Nanomaterials*, 2012, **2012**, Article ID 351205, 16 pages.
[2]   M. Wicklein, S. Ryan, D.M. White and R.A. Clegg, *International Journal of Impact Engineering*, 2008, **35**, 12, 1861.

[3] M. Grujicic, B. Pandurangan, C.L. Zhao, S.B. Biggers and D.R. Morgan, *Applied Surface Science*, 2006, **252**, 14, 5035.

[4] ASTM D7264, *Standard Test Method for Flexural Properties of Polymer Matrix Composite Materials*, 2007.

[5] S. Varghese, A. Lele and R. Mashelkar, *Journal of Polymer Science, Part A: Polymer Chemistry Edition*, 2006, **44**, 1, 666.

[6] J.M. Asua, *Progress in Polymer Science*, 2002, **27**, 7, 1283.

[7] B.J. Blaiszik, M.M. Caruso, D.A. McIlroy, J.S. Moore, S.R. White and N.R. Sottos, *Polymer*, 2009, **50**, 4, 990.

[8] E.B. Murphy and F. Wudl, *Progress in Polymer Science*, 2010, **35**, 1–2, 223.

[9] E.N. Brown, M.R. Kessler, N.R. Sottos and S.R. White, *Journal of Microencapsulation*, 2003, **20**, 6, 719.

[10] X. Liu, X. Sheng, J.K. Lee and M.R. Kessler, *Macromolecular Materials and Engineering*, 2009, **294**, 6–7, 389.

[11] A J. Patel, S.R. White, and E.D. Wetzel, *Self-healing Composite Armor: Self-healing Composites for Mitigation of Impact Damage in US Army Applications*, Final Report, Contract No. W911NF-06-2-0003, US Army Research Laboratory, Adelphi, MD, 2006.

[12] E.N. Brown, S.R. White and N.R. Sottos, *Journal of Materials Science*, 2004, **39**, 5, 1703.

[13] B.J. Blaiszik, N.R. Sottos and S.R. White, *Composites Science and Technology*, 2008, **68**, 3–4, 978.

[14] K.S. Suslick and G.J. Price, *Annual Review of Materials Research*, 1999, **29**, 295.

[15] X. Liu, X. Sheng, J.K. Lee and M.R. Kessler, *Journal of Thermal Analysis and Calorimetry*, 2007, **89**, 2, 453.

[16] ASTM D2344, *Standard Test Method for Short-Beam Strength of Polymer Matrix Composite Materials and Their Laminates*, 2013.

Chapter 9

Mitigating the effect of space small debris on COPV in space with fibre sensors monitoring and self-repairing materials

Small space debris are a high risk for the walls of composite overwrapped pressure vessels (COPV), by making small holes and causing the fuel leak. Commonly the self-healing materials are used to keep the mechanical structure strength; here the hermeticity of the repaired portion is a stringent requirement to prevent any potential fuel leak from the cryogenic tank in vacuum. The efficiency is compared for protective walls composed of a combination of various layers, using strong materials (Kevlar, Nextel) and self-healing commercial materials developed as bulletproof, for example, the ethylene-co-meth acrylic acid (EMAA) and Reverlink[TM].

In this chapter, we review our results regarding the mitigating of the impacts of space small debris on these COPV by using fibre sensors monitoring and self-repairing materials. The small debris impact dynamic was detected and monitored with fibre Bragg gratings (FBG) sensors at very fast acquisition frequencies, up to 500 MHz (2 ns), measuring the variation of the total reflected signal by the FBG. The acquisition system was based on commercially available products. To measure the total wavelength spectrum, the fastest available spectrometer can go up to 2 MHz acquisition (Micron-Optics[TM]), which was judged insufficient to detect the hypervelocity impact. The impact pressure evolution of the FBG, placed in the middle layer of the COPV, was compared with commonly used strain gauges placed a few layers further or on the back of the last layer. The measured impact time delay and relative intensity were compatible between the two sensing methods. Some samples were characterised in details using the X-ray computed tomography at the European Space Research and Technology Centre (ESTEC, Noordwijk, the Nederland), which has permitted to confirm the results by observing the details of the healing and follow the impact trajectory visually.

9.1 Introduction

The work detailed in this chapter aimed to provide a basic solution to protect the COPV walls from the space debris and a way to monitor the debris impact using optical fibre sensors. As depicted in Chapters 7 and 8, the presence in space of micrometeoroids and orbital debris, particularly in low earth orbit, presents a continuous hazard to orbiting spacecrafts such as the space shuttle and the International Space Station (ISS). The mitigation of space meteorites is one of the six major issues for international considera- tion by the United Nations Technical Committee on the Peaceful Uses of Outer Space [1]. The initial impact of debris is a hole, somewhat larger than the debris' diameter, followed by an energetic plume of many hundreds of particles that spread in a defined cone angle to create a damage zone nom- inally one (or many) order of magnitude larger than the impact particle dia- meter. In particular, with composite laminates, a delamination is created internally over a larger area. The requirements are to prevent space debris from creating a hole or heal it and close it hermetically. The hermiticity is verified with a vacuum test. We have monitored the impact evolution and the healing using FBG sensors [2–12]. One wavelength FBGs are commonly used as sensors. Chirped FBGs are sometimes used, as they can have all the periods corresponding to the range of telecom 1520–1565 nm. For the debris impact, the strain was monitored, as the most important effect, while the temperature increase is a much slower process.

9.2   Methodology

The project has included the following steps:

- Select and test self-healing materials,
- Calibrate and use a hypervelocity pellet launcher,
- Select and embed FBG sensors,
- Verify the self-healing hermeticity in vacuum and
- Analyse the healed samples with X-ray computed tomography to follow the pellet trajectory and the healing.

The efficiency of self-healing of COPV was compared for protective walls composed of a combination of various layers, using healing agent embedded in microcapsules, strong materials such as Kevlar and Nextel and self-healing commercial materials developed as bulletproof, for example, the ethylene-co-methacrylic acid (EMAA) and Reverlink^{TM}. The hermeti- city of the repaired portion is a stringent requirement to prevent any potential fuel leak from the cryogenic tank in vacuum. Microcapsules filled with 5-ethylene-2-norboren (5E2N) as those used previously by MPB within composite layers, permitted to repair the small cracks and delami- nation, around the impact hole; however, they were not efficient as a hermetic layer.

The tradeoff led us to select a multilayered mixture of two materials:

- Self-healing polymers such as ionomer-EMAA (ethylene-co-methacrylic acid (bullet-proof materials)) or supra molecule rubber Reverlink^{TM}.
- Strong materials resistant to impact such as Kevlar (polymer-aramid) and Nextel (ceramic-glass).

To experimentally simulate the small debris impact, we have used a double stage launcher (Figure 9.1) kindly provided by McGill University (Montreal, Quebec, Canada) with aluminium and stainless spheres of 2–4 mm diameter, providing pellets hypervelocity in the range of 1–1.7 km/s. The most of the test were made with stainless steel 2 mm in diameter, which has higher effects than the Al pellets of the same diameter at the same velocity, due to the higher density.

Figure 9.2 shows a picture of the pellet/sabot and the diaphragm. The magnet bar is used to measure the hypervelocity when it passes between two copper coils separated by 5 cm.

Two types of FBGs were used, namely:

- A thin central wavelength FBG with 0.5 nm spectral width and 1 cm length similar to those used as wave division multiplexing (WDM) in telecom, it is used as a sensor since the pressure impact proportional to the wavelength shift.
- A chirped FBG covering about 40 nm reflectance range with length about 2.5–4 cm long. The position on the grating and the local wavelength are linearly related. A measurement of the complete spectrum of the FBG sensor before and after the impact gives important information in synergy

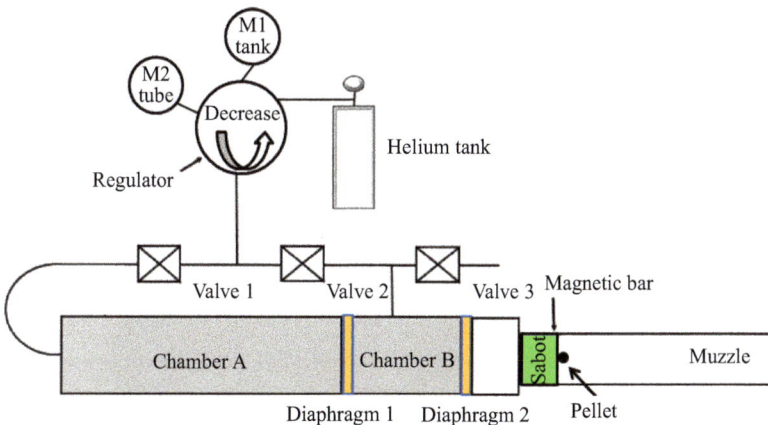

Figure 9.1   Two stages launcher pressure connection (up to 1.7 km/s, pellets: 1–5 mm in diameter)

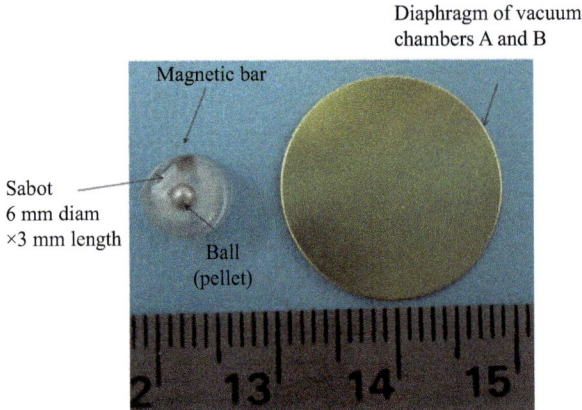

Figure 9.2   Diaphragm and a pellet (ball) with a sabot

with the fast signal measurements. The hypervelocity impact dynamic is monitored with the change of the FBG reflected intensity, in time and value, permitting the measurement of the local stresses and the destroyed region. Moreover, the slow change of the residual strain and self-healing as well as their localisation are obtained by comparing the detailed reflectance spectra of the sensors, before the test, a few minutes after the test, one day after the test and a few weeks later, for slow evolution.

Figure 9.3 illustrates the single wavelength sensor submitted to compression expansion in the direction of the fibre. The shape is conserved even with a shift of about 30 nm. Spectrometers for the FBGs have an acquisition speed limited to 2 MHz (Micron-Optics^{TM}), which is slow to catch the hypervelocity impacts (Figure 9.4).

We proposed to use the total intensity reflected by chirped FBG covering a wide range of wavelength. This intensity after impact is linearly related to the position impacted. The use of total intensity provides a main advantage; it can be measured using faster, simpler and easily available electronic components in GHz levels. More information can be obtained by measuring the complete spectrum after the shot to see the final status. Figures 9.5 and 9.6 illustrate the linear relationship between the total intensity, the physical length and the wavelength in a plateau chirped FBG.

9.3   Experimental results

Representative results were achieved using (Figure 9.7):

- four layers Kevlar and epoxy,
- fibre sensor in the middle and
- strain gauge on the back side.

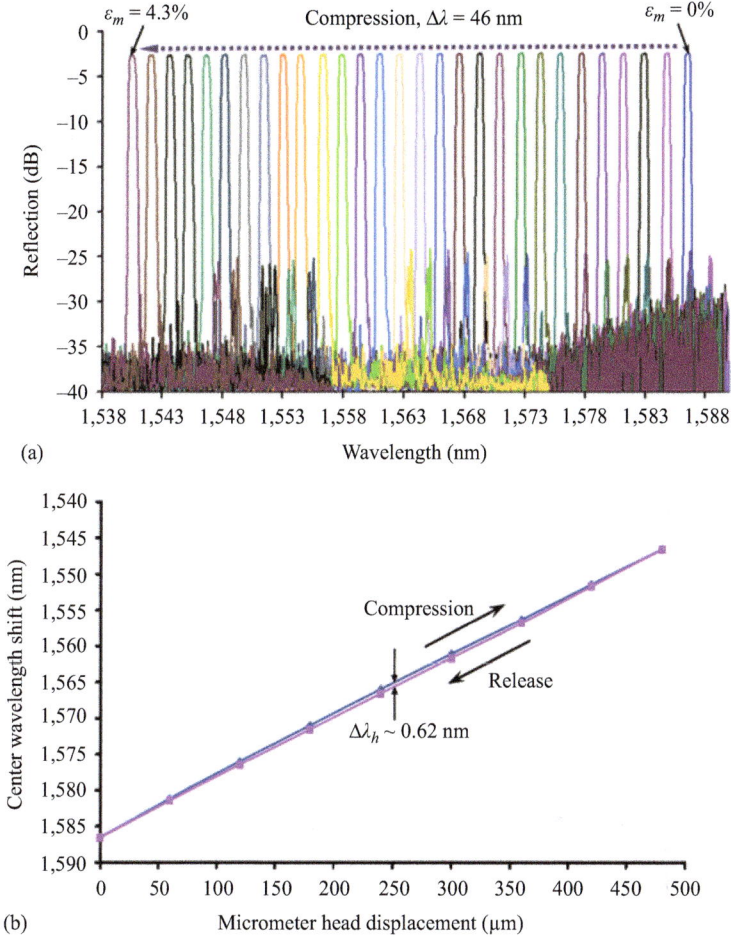

Figure 9.3 (a) Single wavelength sensor submitted to compression expansion in the direction of the fibre and (b) centre wavelength shift and hysteresis with compression–expansion in the direction of the fibre

The debris simulants are 2 mm diameter stainless steel spherical pellets launched at 1.5 km/s.

The measurements are presented in the following figures:

We can see in Figure 9.8 that there is 430 ns delay between the response of the FBG and strain gauge due to the few mm thick layers that the pellet has to pass between the fibre and the strain gauge.

Figures 9.9 and 9.10 show the complete FBG reflected spectra before and after the impact. The impact breaks a small part (1.2 mm) of the fibre, as can be deduced from the missing part of the spectrum. The left and right side show strong stresses after the impact, with some recuperation (healing) after 86 days.

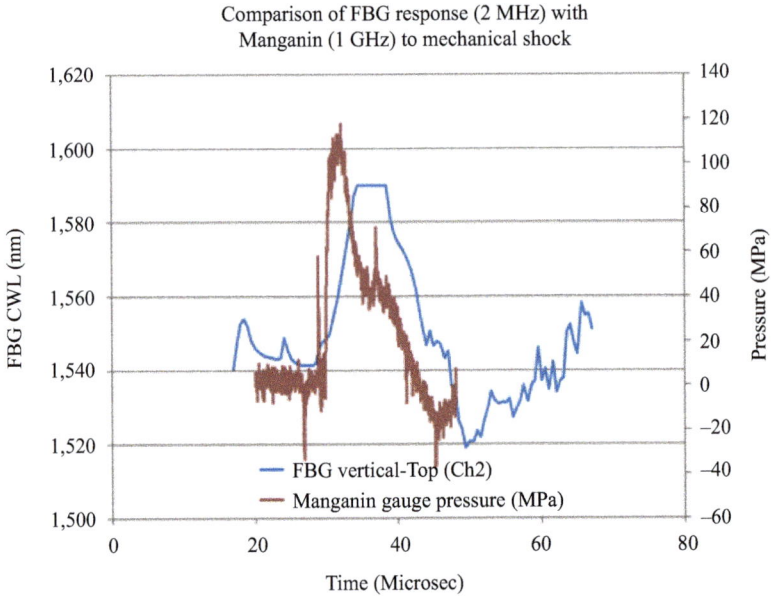

Figure 9.4   *Comparison of the strain gauge and FBG response to mechanical shock*

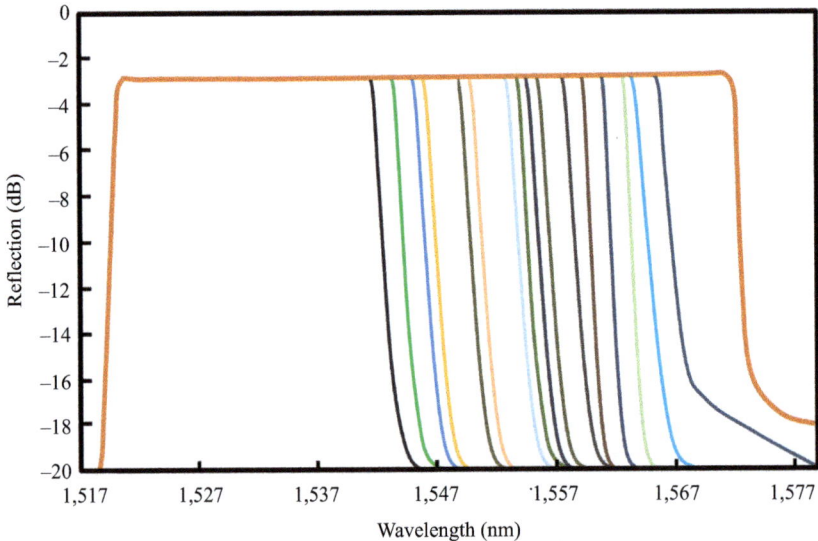

Figure 9.5   *Example of plateau chirped FBG reflection after different cuts, starting at the highest wavelength (all the FBG signals were smoothed by Savitzky-Golay, Percentile Filter method for clarity)*

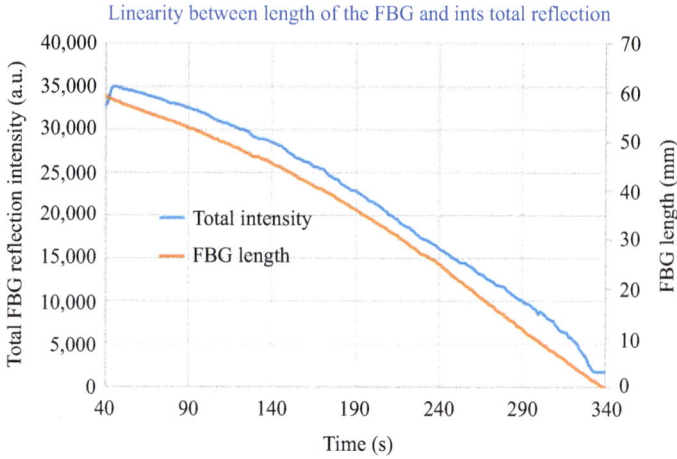

Figure 9.6   Linear relationship wavelength versus its position on the grating and linear intensity versus FBG lost part

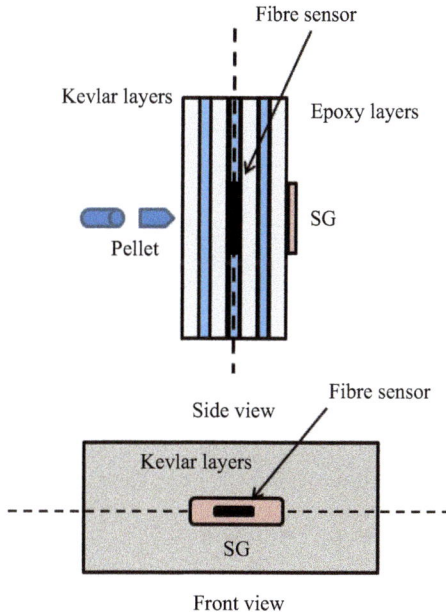

Figure 9.7   Four layers Kevlar/epoxy: (a) cross-section side view strain gauge and FBG (example shot #79) and (b) top front view

The left side has two local residual stress (peak intensity) and the stress is reduced with time.

Figures 9.11 and 9.12 show the results of FBG single wavelength and strain gauge fast acquisition (shot #75). The strain gauge behind the four layers responds 250 ns after the FBG in the middle of the layers.

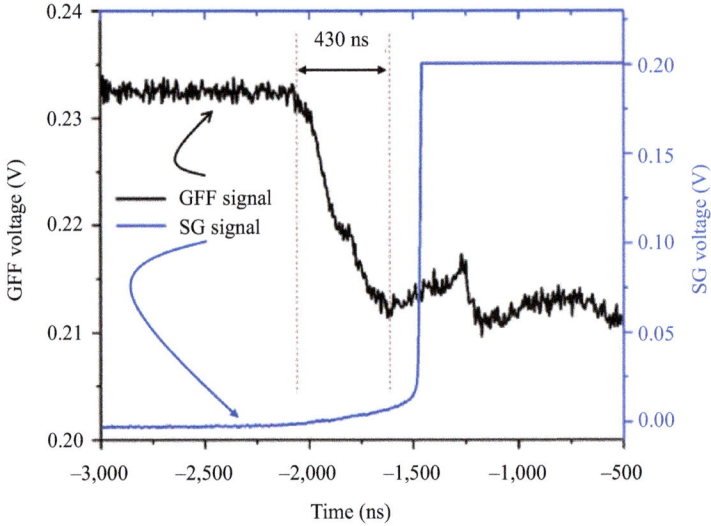

Figure 9.8   Fast response of the FBG (called GFF) and the strain gauge (SG)

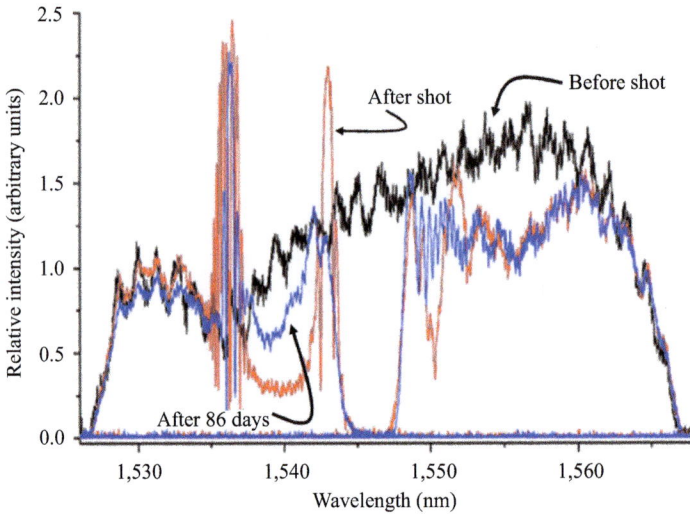

*Figure 9.9   FBG spectra before the impact (black), the empty part in the
middle is the broken part, the red graphics are the FBG spectra
(left and right side) just after the impact and the blue graphics
are the FBG spectra (left and right side) after 86 days*

Table 9.1 shows the shot #75 strain details measured just after the shot
from the multi-CWL shifts.

The physical size of the single wavelength sensor (FBG length) is about
10 mm. We cannot identify where exactly the pellet hit; however, we can
identify the values of a few local strains, without identifying their

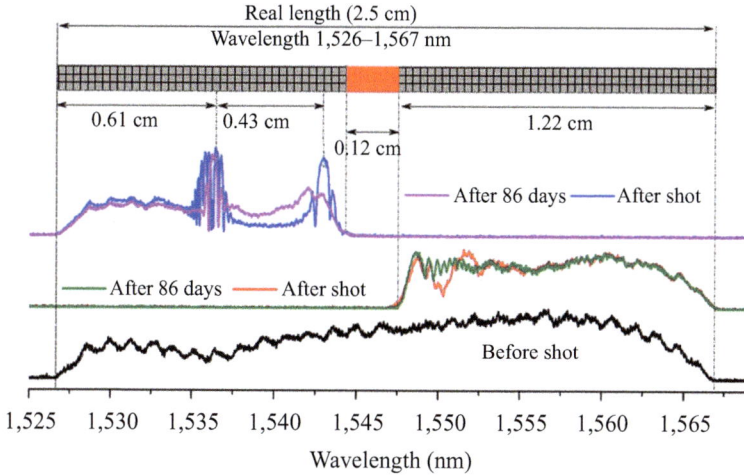

Figure 9.10 *Spectra of the FBG before and after impact traced in cascade for better clarity*

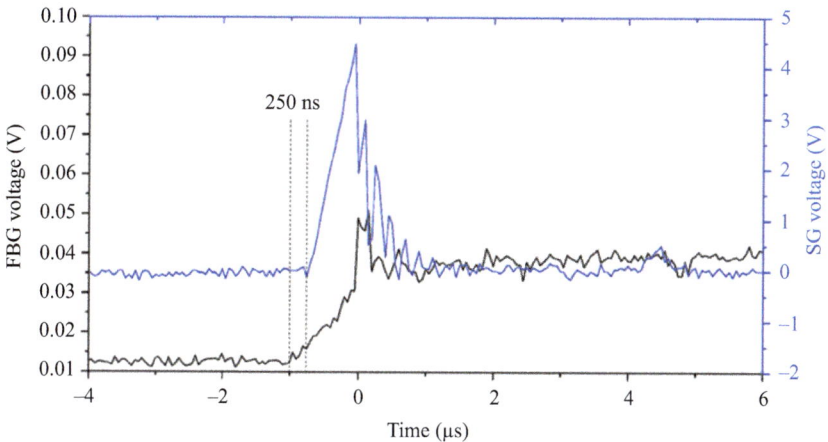

Figure 9.11 *Fast response of the FBG (single wavelength) and the strain gauge (SG)*

position. In the chirped FBG we could identify the local position of the strains; however, without knowing their values, only a qualitative idea is provided with the change of the reflected intensity.

9.4 Healing verification

The healing hermeticity is demonstrated by testing the Kevlar-EMAA multilayers in vacuum. The multilayers were hermetic in vacuum at 10^{-6} Torr. Figure 9.13 shows the detailed SEM images of the impact and healing of the Kevlar-EMAA multilayers.

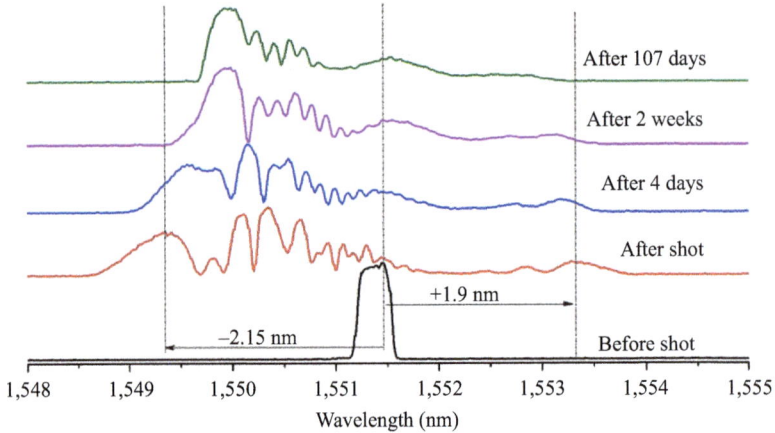

Figure 9.12   Spectra in cascade, of the single wavelength FBG before and after the impact – we can see some recuperation with a permanent induced stress (wavelength shift)

Table 9.1   Strain details measured just after the shot from the multi-CWL shifts (negative shift in wavelength = stress; positive shift in wavelength = extension)

Initial peak (nm)	Shifted peaks (nm)	Pressed Δλ (nm)	Strain Δε (%)	Extended Δλ (nm)	Strain Δε (%)
1,551.4	1,549.3	−2.1	−0.173		
	1,549.8	−1.6	−0.132		
	1,550.1	−1.3	−0.107		
	1,550.3	−1.1	−0.091		
	1,550.6	−0.8	−0.066		
	1,550.9	−0.5	−0.041		
	1,551.0	−0.4	−0.033		
	1,551.3	−0.1	−8.2E−03		
	1,552.4			1	0.0823
	1,552.8			1.4	0.1152
	1,553.3			1.9	0.1564

The X-ray computed tomography measurements were performed at ESTEC/ESA Test Centre. The images were obtained over 360° with micrometric resolution. The scans set up were at 104 kV and 155 μA. Postmeasurements softwares provided by ESTEC-Laboratory are used for assembling all images together in a 3D volume rendering. The X-rays computed tomography image permits to see the EMAA was healed, and to follow the pellet trajectory through the Kevlar that does not heal (Figure 9.14).

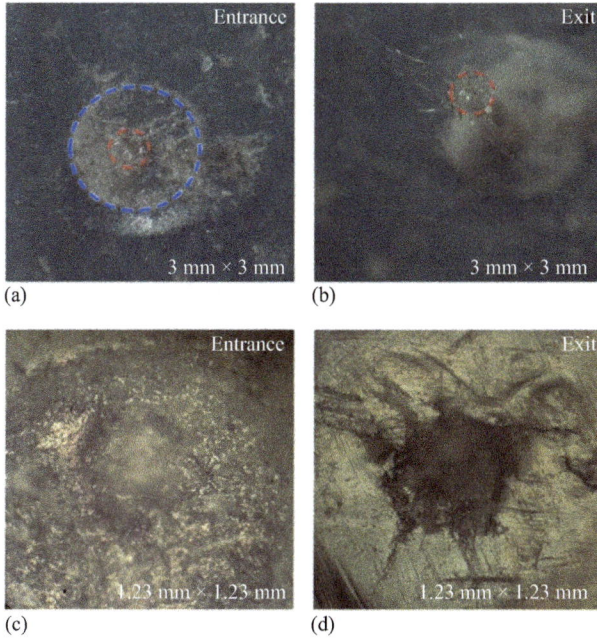

Figure 9.13   (a) The bullet entering side after impact; (b) the bullet exiting side after impact and (c) and (d) are enlargement of (a) and (b). The red circle indicates the influenced area at the entering side. The blue circle indicates the exiting side

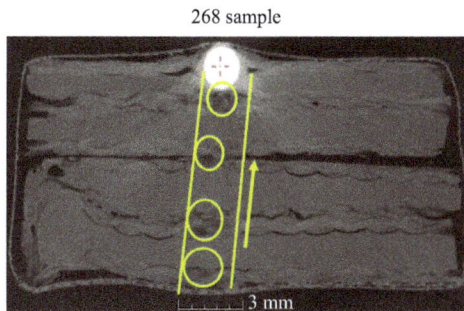

Figure 9.14   X-rays computed tomography image of an impacted Kevlar-EMAA multilayer

9.5   Conclusions

In this chapter, we have investigated the possibility of combining layers of commercial materials which are efficient as self-healing layers of COPV. The combination should contain strong materials layers (Kevlar, Nextel, …) and self-healing materials (e.g., Surlyn EMAA). There are innovative potential applications of fibre sensors to monitor the small debris impact.

Fibre sensors can monitor very fast events at the nanosecond level along with the evolution of impacted material before and after impact at a slow time level. Selecting an optimal combination of materials EMAA, Kevlar/Nextel and Resin would be the best method to protect the COPV tank. The hypervelocity is lower than the majority of space debris velocity. The results discussed in this chapter have demonstrated the feasibility and readiness for the further step to achieve by using higher velocity test and complete cover of the COPV.

References

[1]   http://www.unoosa.org/oosa/en/ourwork/copuos/index.html
[2]   E.W. Taylor, S.J. McKinney, A.D. Sanchez, *et al.*, 'Gamma-ray induced effects in erbium doped fibre optic amplifiers', *Proceedings of SPIE Conference on Photonics for Space Environments VI*, San Diego, CA, USA, 1998, **3440**, 16.
[3]   B.P. Fox, Z.V. Schneider, K. Simmons-Potter, *et al.*, 'Gamma radiation effects in Yb-doped optical fiber', *Proceedings of SPIE Conference on Fibre Lasers IV: Technology, Systems, and Applications*, San Jose, CA, USA, 2007, **6453**, 645328.
[4]   H. Henschel, O. Köhn and U. Weinand, *IEEE Transactions on Nuclear Science*, 2002, **49**, 3, 1401.
[5]   H. Henschel, M. Koerfer, J. Kuhnhenn, U. Weinand and F. Wulf, 'Fibre optic sensor solutions for particle accelerators,' *Proceedings of the SPIE 5855, 17th International Conference on Optical Fibre Sensors*, 23 May 2005, Bruges, Belgium.
[6]   K.V. Zotov, M.E. Likhachev, A.L. Tomashuk, *et al.*, *Proceedings of the 9th European Conference on Radiation and Its Effects on Components and Systems (RADECS)*, 2007, doi:10.1109/RADECS.2007.5205517.
[7]   K.V. Zotov, M.E. Likhachev, A.L. Tomashuk, *et al.*, 'Radiation resistant Er-doped fibers: optimization of pump wavelength', *IEEE Photonics Technology Letters*, 2008, **20**, 17, 1476–1478.
[8]   F. Berghmansa, A.F. Fernandeza, B. Bricharda, *et al.*, *Proceedings of SPIE International Symposium on Industrial and Environmental Monitors and Biosensors Harsh Environment Sensors*, 1998, **3538**, 28.
[9]   A.I. Gusarov, F. Berghmans, O. Deparis, *et al.*, 'High total dose radiation effects on temperature sensing fiber Bragg gratings', *IEEE Photonics Technology Letters*, 1999, **11**, 9, 1159–1161.
[10]   E.W. Taylor, *Proceedings of the 1999 IEEE Aerospace Conference*, 1999, **3**, 307.
[11]   A. Gusarov, D. Kinet, C. Caucheteur, M. Wuilpart and P. Megret, *IEEE Transactions on Nuclear Science*, 2010, **57**, 6, 3775.
[12]   A. Gusarov, B. Brichard and D. Nikogosyan, *IEEE Transactions on Nuclear Science*, 2010, **57**, 4, 2024.

Chapter 10

Conclusions and outlook into the future

We have presented a series of recent results related to the various self-healing concepts and systems. Research of self-healing materials is an active and exciting field, with an increasing number of articles published every year. This research covers a wide spectrum of different materials and methods such as healing of concrete structures using embedded glass fibres and the more recent work on healing using shape memory alloy wires in a polymer composite, and/or the use of a multidimensional microvascular network for the healing applications. These various avenues are being explored with the overall aim of achieving prolonged functional lifetimes for composite structural materials. It is fair to say that the results achieved are astounding.

Because of the strong interest of both academic and commercial researchers in the hollow fibre and microencapsulation approaches to self-healing polymer development, new types of self-healing technology have been emerging at an increasing rate over the last decade. Indeed, in recent years, promising prospectives have opened up for the design of innovative self-healing nanosystems. Computer simulations have provided useful indications that have directed the efforts of scientists towards the fabrication of repair systems. Starting from the idea of repairing composite materials research has spread out over different fields:

- Electronic circuits using liquid metal or small nanotubes,
- Chemical processes using solar energy,
- Graphene material-based self-repair,
- Curing epoxy resins in space for inflatable structures and
- Paint that heals its own scratches once it is subjected to sunlight.

In the composite field, the repairing of small cracks induced artificially has been extended to very wide effects such as

- Fast phenomena (birds hitting an airplane),
- Ultra-fast phenomena (hypervelocity impact tests simulating space debris) protecting structure and
- Very harsh change in environment (thermal shock).

Various healing agents and materials are currently used, including:

- Monomers that polymerise through metathesis, for composites,
- Anti-corrosion microcapsules for metallic surfaces,
- Ceramics-based materials used, for example, as atmospheric re-entry vehicle coatings with high thermal resistant capability,
- Liquid metals for electronic circuits,
- Elastomers that resist bullets and in some cases space debris and
- Carbon nanotubes (CNTs) used as additives to increase the strength of a material.

Looking into the future, as has already been discussed in this book, the durability of materials is probably one of the main challenges encountered today for structural as well as coating applications. Material degradation can occur for a wide variety of reasons, such as fatigue loadings, thermal effects, corrosion, or more generally, for environmental effects of all kinds. Materials failure normally starts at the nanoscale level and is then amplified to the macro scale level until catastrophic failure occurs. Therefore, the ideal solution would be to block and/or eliminate damage as it occurs at the nano/microscale and restore the original material properties.

We have seen that the healing process can be initiated by means of an external source of energy (stimuli), as was shown in the case of bullet penetration [1] where the ballistic impact caused local heating of the material allowing self-healing of ionomers, or in the case of self-healing paint used in the automotive industry. In the latter case, small scratches can be restored by solar heating [2]. Single cracks formed in polymethyl methacrylate specimens at room temperature were also shown to be completely restored above the glass transition temperature [3–5]. The presence of noncovalent hydrogen bonds [6] in mechano-sensitive polymers may allow for a rearrangement of principal chemical bonds, so that they can be used for self-healing. Numerical studies have also shown that nanoscopic gel particles, which are interconnected in a macroscopic network by means of stable and labile bonds, have the potential to be used in self-healing applications.

To date, all the techniques that are currently used are, however, limited by the container size. Containers should be in the nanoscale range since larger ones could lead to large hollow cavities, which could compromise the mechanical properties of the hosting structural material and/or the passive protective properties of the coating material [7]. Besides which, advanced materials are designed to be either tough or self-healing, but typically not both. It would be ideal to have a material, which could be at the same time tougher and self-repairable, but this is still not possible with current technologies.

CNTs are considered to be an ideal filler material for mechanical reinforcement as well as ideal molecular storage devices. This is because CNTs are very small, and thus, they have an extremely large interfacial area. CNTs have interesting mechanical and chemical properties, and have a hollow tubular structure. Polymer/CNT composites [8] have already shown many promising results and various materials such as hydrogen [9], metal and/or metal carbide [10], fullerenes [11], methane [12] and DNA [13] have been successfully inserted inside CNTs. Although a great deal of work has been done with CNTs as self-storage devices, CNTs have not been yet investigated as nanoreservoirs for self-healing applications.

The main challenges for this application are how to insert molecules into the CNTs, whether cracks can form on the sidewall of a CNT during its propagation, and if the healing agent will flow out of the CNT when the crack is formed. Lanzara *et al.* [14] have investigated the use of CNTs as nanoreservoirs for automatic repairing applications, through a molecular dynamics study with particular focus on the CNT capacity of delivering a healing agent. Authors have shown that the CNTs were not only able to carry the catalytic healing agent for local repair, but they can also simultaneously play the role of filler material for mechanical reinforcement prior to and after the delivery of the active material. In this context it is interesting to note the work of Stoffa *et al.* on the self-healing seal for solid oxide fuel cells (SOFCs) [15]. The authors used the yttria-stabilised-zirconia (YSZ) as fillers for making glass-composites as a self-healing seal for SOFC. The YSZ-glass system successfully passed tests at up to 1,000 °C and even higher. Indeed, SOFC should prevent leakage over high temperatures (>800 °C) of fuel and oxidant streams, as well as the reactant. The seal material is required to be electrically isolated and to be mechanically and chemically stable during the reactions. As a matter of fact, a research group from the Los Alamos National Laboratory (USA) demonstrated through simulation that a tungsten layer would facilitate the vacancy-interstitial recombination and thereby reduce the voids into the fusion reactor, in the scenario of damage caused by neutrons [16]. Recently, a group from Japan has developed an erbium oxide insulating coating to self-repair a Li/V-alloy type fusion blanket [17]. From a long-term point of view, synthesis of polymers accompanied by an intrinsic self-healing function through molecular design would be revolutionary. Recent exploration has shown the prospects of this trend, but the automatic trigger mechanism remains to be resolved. Solving these problems will certainly push polymer science and engineering research forward.

References

[1] R.J. Varley and S. Van der Zwaag, *Acta Materialia*, 2008, **56**, 19, 5737.

[2] S. van Der Zwaag, *Self-healing Materials – An Alternative Approach to 20 Centuries Materials Science*, Eds., S. van der Zwaag, Springer, Dordrecht, the Netherlands 2007.

[3] K. Jud, H.H. Kausch and J.G. Williams, *Journal of Materials Science*, 1981, **16**, 1, 204.

[4] H.H. Kausch, and K. Jud, *Plastics and Rubber Processing and Applications*, 1982, **2**, 3, 265.

[5] H.H. Kausch, *Pure and Applied Chemistry*, 1983, **55**, 5, 833.

[6] R.P. Sijbesma, F.H. Beijer, L. Brunsveld, *et al.*, *Science*, 1997, **278**, 5343, 1601.

[7] M. Zako and N. Takano, *Journal of Intelligent Material Systems and Structures*, 1999, **10**, 10, 836.

[8] J.N. Coleman, U. Khan and Y.K. Gun'ko, *Advanced Materials*, 2006, **18**, 6, 689.

[9] C. Liu, Y.Y. Fan, M. Liu, H.T. Cong, H.M. Cheng and M.S. Dresselhaus, *Science*, 1999, **286**, 5442, 1127.

[10] C. Guerret-Piécourt, Y. Le Bouar, A. Lolseau and H. Pascard, *Nature*, 1994, **372**, 6508, 761.

[11] Y. Xue and M. Chen, *Materials Research Society Proceedings*, 2005, **899**, 3.

[12] B. Ni, S.B. Sinnott, P.T. Mikulski and J.A. Harrison, *Physical Review Letters*, 2002, **88**, 20, 205505.

[13] H. Gao, Y. Kong, D. Cui and C.S. Ozkan, *Nano Letters*, 2003, **3**, 4, 471.

[14] G. Lanzara, Y. Yoon, H. Liu, S. Peng and W-I. Lee, *Nanotechnology*, 2009, **20**, 33, 335704.

[15] R.N. Singh, *Innovative Self-healing Seals for Solid Oxide Fuel Cells (SOFFC), Final Report DOE Award # DE-09FE001390*, University of Cincinnati, OH, submitted to U.S. Department of Energy National Energy Technology Laboratory, Pittsburgh, PA, 2012. Available at: https://www.osti.gov/servlets/purl/1054518

[16] V. Borovikov, X. Z. Tang, D. Perez, X. M. Bai, B. P. Uberuaga and A. F. Voter, *Nuclear Fusion*, 2013, **53**, 6, 063001.

[17] Z. Yao, A. Suzuki, T. Muroga and T. Nagasaka, *Fusion Engineering and Design*, 2006, **81**, 23–24, 2887.

Index